U.S. Department
of Transportation
**National Highway
Traffic Safety
Administration**

DOT HS 811 516

October 2011

Integrated Vehicle-Based Safety Systems (IVBSS)

Light Vehicle Field Operational Test Independent Evaluation

DISCLAIMER

This publication is distributed by the U.S. Department of Transportation, National Highway Traffic Safety Administration, in the interest of information exchange. The opinions, findings, and conclusions expressed in this publication are those of the authors and not necessarily those of the Department of Transportation or the National Highway Traffic Safety Administration. The United States Government assumes no liability for its contents or use thereof. If trade names, manufacturers' names, or specific products are mentioned, it is because they are considered essential to the object of the publication and should not be construed as an endorsement. The United States Government does not endorse products or manufacturers.

REPORT DOCUMENTATION PAGE		Form Approved OMB No. 0704-0188
Public reporting burden for this collection of information is estimated to average 1 hour per response, including the time for reviewing instructions, searching existing data sources, gathering and maintaining the data needed, and completing and reviewing the collection of information. Send comments regarding this burden estimate or any other aspect of this collection of information, including suggestions for reducing this burden, to Washington Headquarters Services, Directorate for Information Operations and Reports, 1215 Jefferson Davis Highway, Suite 1204, Arlington, VA 22202-4302, and to the Office of Management and Budget, Paperwork Reduction Project (0704-0188), Washington, DC 20503.		
1. AGENCY USE ONLY (Leave blank) DOT HS 811 516	2. REPORT DATE October 2011	3. REPORT TYPE AND DATES COVERED Final Report August 2006-December 2010
4. TITLE AND SUBTITLE Integrated Vehicle-Based Safety Systems (IVBSS) Light Vehicle Field Operational Test Independent Evaluation		5. FUNDING NUMBERS Inter-Agency Agreement HS-22A1 DTNH22-08-V-00015
6. AUTHOR(S) Emily Nodine, Andy Lam, Scott Stevens, Michael Razo, and Wassim Najm		
7. PERFORMING ORGANIZATION NAME(S) AND ADDRESS(ES) U.S. Department of Transportation Research and Innovative Technology Administration John A. Volpe National Transportation Systems Center Cambridge, MA 02142		8. PERFORMING ORGANIZATION REPORT NUMBER DOT-VNTSC-NHTSA-11-02
9. SPONSORING/MONITORING AGENCY NAME(S) AND ADDRESS(ES) Raymond J. Resendes U.S. Department of Transportation National Highway Traffic Safety Administration		10. SPONSORING/MONITORING AGENCY REPORT NUMBER DOT HS 811 516
11. SUPPLEMENTARY NOTES		
12a. DISTRIBUTION/AVAILABILITY STATEMENT Document is available to the public through the National Technical Information Service www.ntisgov		12b. DISTRIBUTION CODE
13. ABSTRACT (Maximum 200 words) This report presents the methodology and results of the independent evaluation of a prototype integrated crash warning system for light vehicles as part of the Integrated Vehicle-Based Safety Systems initiative of the United States Department of Transportation's Intelligent Transportation System program. The system integrates rear-end crash, curve-speed warning, lane change crash, and lane departure warning functions. The goals of the independent evaluation are to assess the safety impact, gauge driver acceptance, and characterize the capability of the integrated safety system. The evaluation is based on naturalistic driving data collected from a field operational test using 108 subjects who drove 16 passenger vehicles equipped with a prototype integrated safety system and a data acquisition system. The test subjects accumulated over 213,000 miles during a 12-month period throughout parts of southeast Michigan. For each driver, the test period was divided into a 12 day baseline condition with the system disabled and a 28 day treatment condition with the system enabled to compare the effect of the system on driving performance. The results of the analysis suggest that driving with the integrated safety system improves driver behavior and increases driver safety, that drivers feel that the system provides a safety benefit, and that the system alerts had a high degree of accuracy. This report delineates the methodology of the different analyses and discusses their results.		
14. SUBJECT TERMS Vehicle safety, crash warning, driver-vehicle interface, intelligent vehicles, driving conflicts, driver acceptance, driving performance measures, Integrated Vehicle-Based Safety System (IVBSS), near crashes, system capability		15. NUMBER OF PAGES 156
		16. PRICE CODE
17. SECURITY CLASSIFICATION OF REPORT Unclassified	18. SECURITY CLASSIFICATION OF THIS PAGE Unclassified	19. SECURITY CLASSIFICATION OF ABSTRACT Unclassified	20. LIMITATION OF ABSTRACT Unlimited

NSN 7540-01-280-5500

Standard Form 298 (Rev. 2-89)
Prescribed by ANSI Std. 239-18
298-102

METRIC/ENGLISH CONVERSION FACTORS

ENGLISH TO METRIC

LENGTH (APPROXIMATE)
- 1 inch (in) = 2.5 centimeters (cm)
- 1 foot (ft) = 30 centimeters (cm)
- 1 yard (yd) = 0.9 meter (m)
- 1 mile (mi) = 1.6 kilometers (km)

AREA (APPROXIMATE)
- 1 square inch (sq in, in^2) = 6.5 square centimeters (cm^2)
- 1 square foot (sq ft, ft^2) = 0.09 square meter (m^2)
- 1 square yard (sq yd, yd^2) = 0.8 square meter (m^2)
- 1 square mile (sq mi, mi^2) = 2.6 square kilometers (km^2)
- 1 acre = 0.4 hectare (he) = 4,000 square meters (m^2)

MASS - WEIGHT (APPROXIMATE)
- 1 ounce (oz) = 28 grams (gm)
- 1 pound (lb) = 0.45 kilogram (kg)
- 1 short ton = 2,000 pounds (lb) = 0.9 tonne (t)

VOLUME (APPROXIMATE)
- 1 teaspoon (tsp) = 5 milliliters (ml)
- 1 tablespoon (tbsp) = 15 milliliters (ml)
- 1 fluid ounce (fl oz) = 30 milliliters (ml)
- 1 cup (c) = 0.24 liter (l)
- 1 pint (pt) = 0.47 liter (l)
- 1 quart (qt) = 0.96 liter (l)
- 1 gallon (gal) = 3.8 liters (l)
- 1 cubic foot (cu ft, ft^3) = 0.03 cubic meter (m^3)
- 1 cubic yard (cu yd, yd^3) = 0.76 cubic meter (m^3)

TEMPERATURE (EXACT)
[(x-32)(5/9)] °F = y °C

METRIC TO ENGLISH

LENGTH (APPROXIMATE)
- 1 millimeter (mm) = 0.04 inch (in)
- 1 centimeter (cm) = 0.4 inch (in)
- 1 meter (m) = 3.3 feet (ft)
- 1 meter (m) = 1.1 yards (yd)
- 1 kilometer (km) = 0.6 mile (mi)

AREA (APPROXIMATE)
- 1 square centimeter (cm^2) = 0.16 square inch (sq in, in^2)
- 1 square meter (m^2) = 1.2 square yards (sq yd, yd^2)
- 1 square kilometer (km^2) = 0.4 square mile (sq mi, mi^2)
- 10,000 square meters (m^2) = 1 hectare (ha) = 2.5 acres

MASS - WEIGHT (APPROXIMATE)
- 1 gram (gm) = 0.036 ounce (oz)
- 1 kilogram (kg) = 2.2 pounds (lb)
- 1 tonne (t) = 1,000 kilograms (kg)
- = 1.1 short tons

VOLUME (APPROXIMATE)
- 1 milliliter (ml) = 0.03 fluid ounce (fl oz)
- 1 liter (l) = 2.1 pints (pt)
- 1 liter (l) = 1.06 quarts (qt)
- 1 liter (l) = 0.26 gallon (gal)
- 1 cubic meter (m^3) = 36 cubic feet (cu ft, ft^3)
- 1 cubic meter (m^3) = 1.3 cubic yards (cu yd, yd^3)

TEMPERATURE (EXACT)
[(9/5) y + 32] °C = x °F

QUICK INCH - CENTIMETER LENGTH CONVERSION

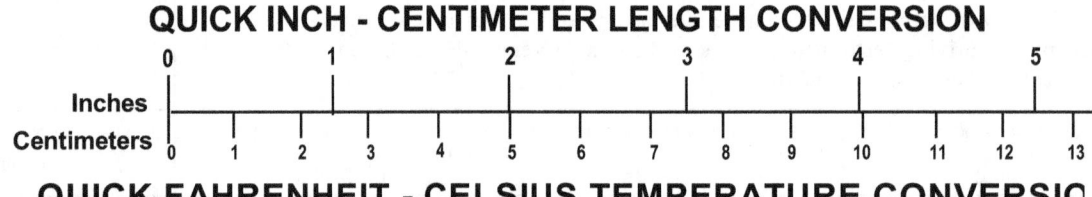

QUICK FAHRENHEIT - CELSIUS TEMPERATURE CONVERSION

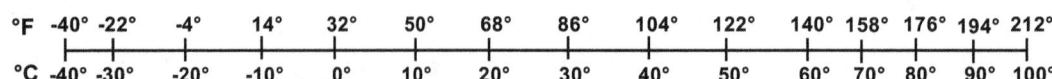

For more exact and or other conversion factors, see NIST Miscellaneous Publication 286, Units of Weights and Measures. Price $2.50 SD Catalog No. C13 10286

Updated 6/17/98

Table of Contents

List of Figures v
List of Tables vii
List of Acronyms ix
Executive Summary 1
1. **Introduction** 4
    1.1 Integrated Safety System 4
    1.2 Target Crashes 5
    1.3 Field Operational Test 6
        1.3.1 Demographics of Field Test Participants 7
        1.3.2 Summary of Field Test Exposure 9
    1.4 Independent Evaluation 11
        1.4.1 Data Processing 11
        1.4.2 Multimedia Data Analysis Tool 13
2. **Safety Impact** 15
    2.1 Safety Impact Technical Approach 15
    2.2 Overall Driving Experience 17
        2.2.1 Speed Maintenance 18
        2.2.2 Headway Keeping 18
        2.2.3 Lane-Change Behavior 18
        2.2.4 Lane Keeping 20
        2.2.5 Vehicle Speed at Curve Entry 22
        2.2.6 Attention to Primary Driving Task 22
    2.3 Conflict Exposure Rates and Driver Response 27
        2.3.1 Overall Driving Conflict Rate 28
        2.3.2 Rear-End Driving Conflicts 28
        2.3.3 Lane-Change Driving Conflicts 29
        2.3.4 Road-Departure driving Conflicts 32
        2.3.5 Curve-Speed Driving Conflicts 36
        2.3.6 Driver Attention in Driving Conflicts 37
    2.4 Near Crash Experiences 38
        2.4.1 Exposure to Near Crashes 39
        2.4.2 Driver Attention in Near Crashes 42
    2.5 Projection of Potential Safety Benefits 43
3. **Driver Acceptance** 46
    3.1 Driver Acceptance Technical Approach 46
        3.1.1 Acceptance by Driver and Objective 46
        3.1.2 Acceptance by Independent Variables 47
    3.2 Subjective Results 48
        3.2.1 General Feedback 48
        3.2.2 Ease of Use 51
        3.2.3 Perceived Usefulness 56
        3.2.4 Ease of Learning 61
        3.2.5 Advocacy 61

        3.2.6   Driving Performance 62
    3.3   Driver Acceptance by Driver Experience Variables 63
        3.3.1   Alert Rate 64
        3.3.2   Driving Patterns 64
        3.3.3   Conflict Rates 66
        3.3.4   False Alarm Rates 67
4. **System Capability** 70
    4.1   Sensors 70
        4.1.1   Forward-Looking Sensors 70
        4.1.2   GPS and Map data 74
        4.1.3   Side-Looking Sensors 74
        4.1.4   Lane Tracking 76
    4.2   Warning Logic 78
        4.2.1   Hazard Propensity 79
        4.2.2   Driver Response 81
        4.2.3   Comparison of Alert Rates between Baseline and Treatment 85
    4.3   Driver-Vehicle Interface 87
    4.4   System Robustness 88
5. **Conclusions** 91
6. **References** 94
Appendix A: Post-Drive Survey 95
Appendix B: Data Processing and Data Mining 110
Appendix C: Video Analysis 113
Appendix D: Video Coding Manual 114
Appendix E: Overall Driving Analysis Supplemental Data 127
Appendix F: Conflict Identification Thresholds 129
Appendix G: Driving Conflict Analysis Supplemental Data 131
Appendix H: Near Crash Thresholds by Conflict Type 142
Appendix I: Post-Drive Survey Mapping to Acceptance Objectives 144

List of Figures

Figure 1. Driver-vehicle interface of the integrated safety system 5
Figure 2. Average age of drivers by age group 8
Figure 3. Highest level of education by age group 9
Figure 4. Number of years with a driver's license by age group 9
Figure 5. Data processing procedures 12
Figure 6. Screen view of multimedia data analysis tool 14
Figure 7. Safety benefits framework 16
Figure 8. Most frequent secondary tasks exhibited during the field test 23
Figure 9. Most frequent secondary tasks exhibited during the field test, by age group 24
Figure 10. Percent of alerts with secondary tasks by age, gender group, and treatment period ... 25
Figure 11. Percent of eyes off forward scene by treatment period by age and gender group 26
Figure 12. Proportion of valid alerted conflicts with secondary tasks by conflict type 38
Figure 13. Proportion of valid alerted conflicts with drivers' eyes off forward scene 38
Figure 14. Proportion of near crashes where drivers were engaged in secondary tasks 42
Figure 15. Proportion of near crashes where drivers had their eyes off the forward scene 43
Figure 16. Average system effectiveness values in various near crashes 44
Figure 17. System features liked best by drivers 49
Figure 18. System characteristics liked least by drivers 50
Figure 19. Situations in which drivers found the integrated system to be most helpful 50
Figure 20. Distribution of ease-of-use responses by age group 51
Figure 21. Distribution of ease-of-use responses by gender 51
Figure 22. Responses to the statement, "the integrated system made driving easier" 52
Figure 23. Responses to the statement, "the integrated system made driving easier" by
 years of driving experience 53
Figure 24. Responses to the statement, "the integrated system was predictable
 and consistent" 53
Figure 25. Drivers' understanding of the warnings by warning type 54
Figure 26. Understanding of warnings by warning type and education level 55
Figure 27. Drivers' responses to the survey item, "The alerts were not annoying" 55
Figure 28. Annoyance with warnings by warning type and education level 56
Figure 29. Distribution of perceived usefulness by age group 57
Figure 30. Distribution of perceived usefulness by gender 57
Figure 31. Drivers' opinions about overall usefulness of the integrated system 58
Figure 32. Drivers' opinions about the safety impact of the integrated system 58
Figure 33. Drivers' reported experience with nuisance warnings 59
Figure 34. Responses about frequency and annoyance with nuisance alerts 60
Figure 35. Reported annoyance with nuisance warnings by driver experience 60
Figure 36. Drivers' willingness to drive with the integrated system 61
Figure 37. Reported changes in driving behavior due to driving with the integrated system 62
Figure 38. Reported reliance on the integrated system 63
Figure 39. Questionnaire responses by overall alert rate 64
Figure 40. Drivers' responses to the statement, "the integrated system made driving easier" 65
Figure 41. Questionnaire responses broken down by drivers' proportion of freeway driving 66

v

Figure 42. Drivers' responses to the questionnaire item, "I always understood why the system provided me with a warning" 66
Figure 43. Responses to questionnaire items relating to drivers' exposure to nuisance alerts 67
Figure 44. Responses to the statement, "Overall I received nuisance warnings (1=Too frequently, 7= Never)" by rate of false LDW-C warnings 68
Figure 45. Responses to the statements, "the system gave me warnings when I did not need them" and "overall, I received nuisance warnings… (1= Too frequently, 7= Never)" by overall false alert rate 69
Figure 46. Proportion of in-path targets for FCW alerts by target type 71
Figure 47. Proportion of in-path targets for FCW alerts by target type and speed bin 71
Figure 48. Target type of alerts issued for stopped out-of-path targets, by speed bin 72
Figure 49. Probability of out-of-path stationary object alert, by vehicle position 73
Figure 50. Range of stationary targets triggering out-of-path FCW alerts on straight roads 73
Figure 51. Curve-speed alerts by road type 74
Figure 52. Target location of LCM alerts 75
Figure 53. Target location of LDW-I alerts 76
Figure 54. Target type of side-imminent alerts issued for non-adjacent targets 76
Figure 55. Lane excursion scenario of LDW-C alerts by road type 77
Figure 56. No excursion LDW-C probability by environmental factors 78
Figure 57. Mapping of valid alerts to driving conflicts 80
Figure 58. Mapping of valid near crashes to alerts 81
Figure 59. Breakdown of driver action in response to valid alerts 82
Figure 60. Average brake reaction time to FCW alerts in baseline and treatment 83
Figure 61. Average peak deceleration response CSW alerts in baseline and treatment 83
Figure 62. Average maximum lateral speed after lateral alerts 84
Figure 63. Average peak lateral acceleration after lateral alerts 85
Figure 64. Drivers' reported usefulness of the physical driver-vehicle interface of the system .. 88
Figure 65. Drivers' opinions about the attention-getting capability of different elements of the driver-vehicle interface 88
Figure 66. Availability of LDW function by travel speed 89
Figure 67. Block diagram of longitudinal driving conflicts 111
Figure 68. Block diagram of lateral driving conflicts 112

List of Tables

Table 1. Annual frequency of target crashes by pre-crash scenario 6
Table 2. Demographics of field test participants 8
Table 3. Exposure of test subjects in the field test 10
Table 4. Alert rates (number of alerts per 100 miles driven) experienced during the field test ... 10
Table 5. Light vehicle video sampling rates 12
Table 6. Number of alerts included in video analysis by alert type 14
Table 7. Results of baseline versus treatment paired t-test for average speed in m/s 18
Table 8. Results of baseline versus treatment paired t-test for mean headway in seconds 19
Table 9. Results for overall number of lane changes per 100 miles driven 19
Table 10. Results for ratio of signaled lane changes to total lane changes 20
Table 11. Results of baseline versus treatment paired t-test for lane busts per
    100 miles driven 21
Table 12. Results of paired t-test for lane bust duration in seconds 22
Table 13. Paired t-test for percent of analyzed alerts with secondary tasks 25
Table 14. Results of paired t-test for percent of analyzed alerts with eyes off forward scene 26
Table 15. Measures used to quantify driver response 28
Table 16. Average number of lane-change conflicts per 100 miles driven 31
Table 17. Average lane incursion time for lane-change conflicts (seconds) 32
Table 18. Average number of conflicts departing curved roads per 100 miles driven 34
Table 19. Average lane bust time for straight road departures (seconds) 35
Table 20. Average maximum lateral acceleration (m/s^2) in curve-speed conflicts 37
Table 21. Breakdown of near crashes and their validity rate 39
Table 22. Paired t-test results of average number of near crashes per 1,000 miles driven 40
Table 23. Paired t-test results of near-crash rates by speed bin and near crash type 41
Table 24. Paired t-test results of road-departure near-crash rates by direction and age/gender
    groups 41
Table 25. Crash reduction estimates with full deployment of the integrated system in light
    vehicles 45
Table 26. Driver experience categories used in driver-acceptance analysis 48
Table 27. Analysis of system alerts 79
Table 28. Average number of alerts per 100 miles driven by treatment period 85
Table 29. Average number of LDW-I alerts per 100 miles driven by treatment period 86
Table 30. Average number of LDW-C alerts per 100 miles driven by treatment period 86
Table 31. Average number of LDW-C alerts per 100 miles driven by departure direction 87
Table 32. Availability of LDW function by travel speed and driving conditions 90
Table 33. Data mining variables 112
Table 34. Breakdown of analyzed alert videos 113
Table 35. Variables coded in video analysis by alert type 113
Table 36: Lane busts per 100 miles driven 127
Table 37: Lane-bust duration (sec) 127
Table 38: Average vehicle speed (m/s) three seconds prior to curve start 128
Table 39. Overall number of conflicts per 100 miles driven 131
Table 40. Average number of rear-end conflicts per 100 miles driven 131
Table 41. Average response intensities to rear-end conflicts 132

Table 42. Average response intensities to lane-change conflicts 136
Table 43. Average number of road-departure conflicts per 100 miles driven 137
Table 44. Average response intensities to road-departure conflicts 138
Table 45. Average number of conflicts where drivers approached curves at excessive
   speed per 100 miles driven 141
Table 46. Average delta speed at CPOI during curve-speed conflicts 141
Table 47. Near crash thresholds by conflict type... 142

List of Acronyms

B	Baseline
BSM	Blind spot monitor
CES	Curve with excessive Speed
CDL	Curved road departure to the left
CDR	Curved road departure to the right
CPOI	Curvature point of interest
CS	Curve speed
CSW	Curve-speed warning
DAS	Data acquisition system
DVI	Driver-vehicle interface
ER	Exposure ratio
FCW	Forward crash warning
FOT	Field operational test
GES	General estimates system
IVBSS	Integrated Vehicle-Based Safety Systems
LCL	Lane change to the left
LCM	Lane-change/merge
LCR	Lane change to the right
LDW	Lane-departure warning
LDW-C	Cautionary lane-departure warning
LDW-I	Imminent lane-departure warning
LVD	Lead vehicle decelerating
LVM	Lead vehicle moving
LVS	Lead vehicle stopped
MDAT	Multimedia data analysis tool
PR	Prevention ratio
RD	Road departure
SDL	Straight road departure to the left
SDR	Straight road departure to the right
SQL	Structured query language
T_{all}	All of the treatment period
T_1	First period of treatment condition
T_2	Second period of treatment condition
TL	Turning to left
TR	Turning to right
UMTRI	University of Michigan Transportation Research Institute

Executive Summary

Background

The objective of the Integrated Vehicle-Based Safety System (IVBSS) initiative is to assess the safety benefits and driver acceptance of a prototype crash warning system for light vehicles (i.e., passenger vehicles under 10,000 lbs, including cars, trucks, vans, etc.). The integrated system includes the following types of crash-imminent warnings:

- Forward-crash warning (FCW)
- Curve-speed warning (CSW)
- Lane-change/merge warning (LCM)
- Lane-departure warning (LDW)
    - LDW cautionary (LDW-C): refers to alerts issued when the vehicle is drifting out of its lane into a clear area (unoccupied lane or clear shoulder).
    - LDW imminent (LDW-I): refers to alerts issued when the vehicle is drifting into an occupied lane or towards a roadside object, causing potential for a collision.

Based on the average of 2004-2008 General Estimates System crash statistics, these four warning functions have the potential to address about 2,674,000 police-reported crashes involving light vehicles annually. The IVBSS initiative is part of the United States Department of Transportation's Intelligent Transportation System program and was led by the University of Michigan Transportation Research Institute. Visteon Corporation, with support from Takata Corporation and Honda R&D Americas, served as the lead developer of the prototype system. The field test discussed in this document was conducted in southeast Michigan. This report presents the results of the independent evaluation of IVBSS for light vehicles, performed by the Volpe National Transportation Center.

Evaluation Goals

The goals of the evaluation were to:
- **Achieve a detailed understanding of system safety benefits***:* this goal estimates the number of crashes that might be avoided by the full deployment of the integrated safety system in the light vehicle fleet in the United States. This goal also addresses unintended consequences in terms of modifications in driver behavior that can have negative side effects on traffic safety.
- **Determine driver acceptance of the system:** this goal assesses the ease of use, perceived usefulness, ease of learning, drivers' advocacy, and drivers' assessment of their own driving performance with the integrated safety system.
- **Characterize system performance:** this goal examines the operational performance of the integrated safety system and its components in the driving environment.

Procedure

The evaluation is based on naturalistic driving data collected from a field operational test using 108 subjects who drove 16 model year 2006 and 2007 Honda Accords with a prototype integrated safety system and a data acquisition system. The subject pool is balanced for age (20-30, 40-50, and 60-70 years old) and gender. Baseline data were collected in the first 12 days of each driver's participation while the last 28 days were dedicated to data collection on driver performance with the system enabled. The analysis was performed on data collected from 68,898 miles driven during the baseline period and 144,496 miles driven during the treatment period. In addition to numerical data analysis, 16,915 videos corresponding to system alerts were analyzed and coded. These alerts included all imminent alerts issued during the field test (FCW, CSW, LCM, and LDW-I), and a random sample of each driver's cautionary drift alerts (LDW-C).

Safety impact was determined through the objective analysis of driver behavior, and the rate of conflict scenarios (driving scenarios in which, had the driver not intervened, a crash would have occurred) near-crash driving scenarios experienced by drivers during the field test. Driver acceptance was assessed through subjective feedback provided by the drivers, and system performance was measured in terms of accuracy of alerts and driver's responses to alerts.

Results

Safety Impact:
- If all passenger cars in the United States were equipped with the integrated safety system, it is estimated that between 162,000 and 788,000 light vehicle crashes could be reduced annually.
- The integrated system showed 40 percent effectiveness in reducing lane-change near crashes and 13 percent effectiveness in reducing road-departure near crashes.
- Drivers showed a significant increase in turn signal usage when making lane changes.
- Drivers showed a 21 percent decrease in the rate of lane busts with the system enabled, indicating improved lane keeping when driving with the system.
- For speeds over 55 mph, there was an overall decrease in conflict rate with the system enabled.
- The rate of lane-change conflicts and road-departure conflicts on curved roads decreased overall with the system enabled.
- Fourteen of the 31 drivers who attended focus groups said that the integrated system helped prevent them from getting into a crash or near crash.
- Drivers did not show an increase in either secondary tasks or eyes-off-forward scene behavior with the system enabled, suggesting that the system did not have unintended consequences with respect to driver attention.
- All drivers showed a reduction in lane-change and road-departure near crashes with the system enabled; the rate of LCM near crashes decreased more for men, and the rate of road departure conflicts decreased more for women.

- Younger drivers showed a 19 percent reduction in all near crashes with the system enabled.

Driver Acceptance:
- Eighty-two percent of drivers felt that the system increased their driving safety
- One third of drivers said that the integrated system issued nuisance warnings too frequently. Younger drivers were less tolerant of the nuisance warnings than middle-aged and older drivers; they were more likely to report that they received too many nuisance warnings and more likely to find the nuisance warnings annoying.
- Drivers found the lateral warning systems to be more useful and more desirable than the longitudinal warnings.
- Drivers reported exposure to false warnings was consistent with their actual exposure.

System Performance:
- Overall, system alerts had a very high degree of accuracy.
- Alerts issued for forward stationary targets were issued mostly for out-of-path targets, indicating a low degree of accuracy for this type of FCW warning.
- Drivers respond to forward threats more quickly and more assertively when they received FCW alerts.
- Drivers showed more deceleration when approaching curves with the system enabled.
- When the system is enabled drivers make more assertive steering responses to resume their lane position after drifting out of their lane.
- Drivers maintained better lane positioning with the system enabled (reduction in LDW warnings).
- With the system enabled, drivers showed a 46 percent reduction in drifts to the left, the type of drift that can lead to a head on collision.

Conclusions

The data showed that the system improved driving performance, decreased exposure to both conflict and near crash driving scenarios, and increased overall driving awareness. Additionally, drivers enjoyed driving with the system and felt that it increased their driving safety. Drivers found the system to be easy to use and easy to understand. Overall, the system issued warnings accurately. However, some warnings may have been issued conservatively, as a low rate of response to warnings was observed.

1. Introduction

This report presents the analytical approach and results of the independent evaluation of a prototype integrated crash warning system for light vehicles (i.e., passenger cars, vans and minivans, sport utility vehicles, and light pickup trucks with gross vehicle weight rating of 10,000 pounds or less). The evaluation is based on naturalistic driving data collected from 16 late model Honda Accord EX vehicles equipped with the prototype integrated safety system. The analytical methods used in the evaluation are outlined, and results are presented and discussed.

1.1 Integrated Safety System

The integrated safety system for light vehicles provides information to assist drivers in avoiding or reducing the severity of crashes through the following four crash warning functions (Sayer et al., 2009):

> **HIGHLIGHTS**
> - The safety system combines rear-end, curve-speed, lane-change, and lane-departure crash warning functions that address approximately 2,674,000 police-reported crashes involving light vehicles annually.
> - 108 drivers of various ages accumulated over 213,000 miles driving 16 Honda Accords equipped with the integrated safety system over a 12-month period.
> - Approximately 12,000 alerts were issued during the treatment period while the system was enabled.

- Forward crash warning (FCW)
- Curve-speed warning (CSW)
- Lane-change/merge (LCM) warning
- Lane-departure warning (LDW)
    - LDW cautionary (LDW-C): refers to alerts issued when the vehicle is drifting out of its lane into a clear area (unoccupied lane or clear shoulder)
    - LDW imminent (LDW-I): refers to alerts issued when the vehicle is drifting into an occupied lane or towards a roadside object, causing potential for a collision

The integrated system addresses crashes in which an equipped vehicle strikes the rear end of another vehicle (FCW), approaches a curve at excessive speed (CSW), changes lanes or merges into traffic and collides with another same-direction vehicle (LCM), and unintentionally drifts off the road edge or crosses a lane boundary (LDW).

The driver-vehicle interface (DVI) consists of visual and audio alerts that warn the driver of the occurrence of one of the above situations. Each alert has a unique audio tone and message displayed on a center display to assist the driver in understanding which type of threat is present. Audible alerts are delivered through the driver headrest speakers, with right-left directionality. The vehicle is also equipped with blind spot monitors (BSM) to help increase driver awareness of objects that are in the driver's blind spot. LDW-C alerts are unique in that they do not issue an audio tone to the driver, but rather provide a haptic warning via vibrations in the side of the driver's seat in which the threat is present. The FCW alert is also augmented with a haptic brake pulse. Figure 1 shows the elements of the driver-vehicle interface.

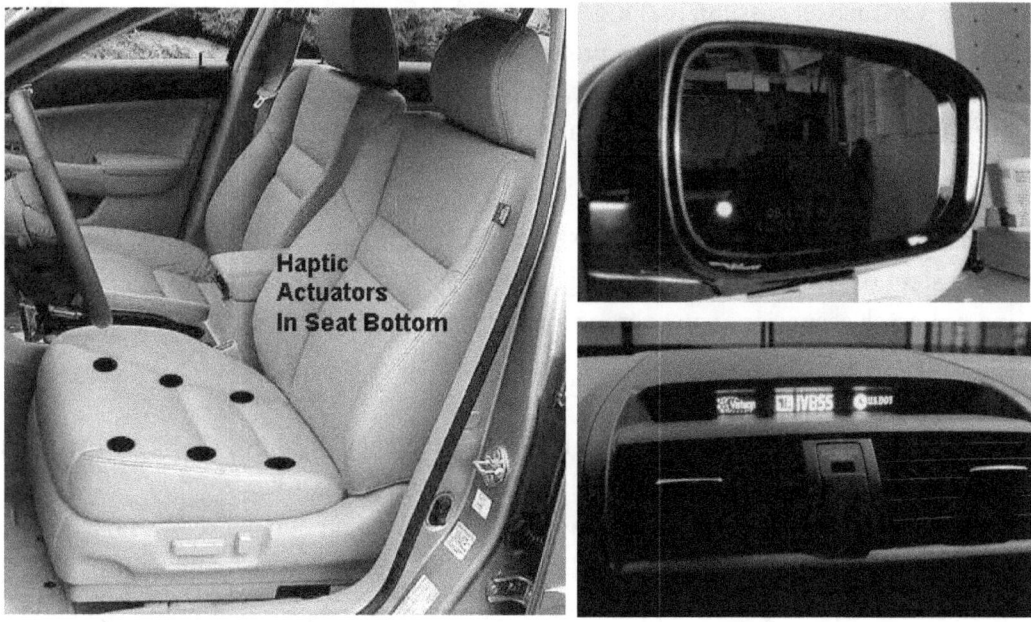

Figure 1. Driver-vehicle interface of the integrated safety system

1.2 Target Crashes

The integrated safety system was designed to address the following pre-crash scenarios, which identify vehicle movements and the critical event prior to a crash (Najm et al., 2007):

- Rear-end—lead vehicle moving (LVM): driver is going straight or decelerating and then closes in on a lead vehicle moving at a lower constant speed.
- Rear-end—lead vehicle decelerating (LVD): driver is going straight while following another lead vehicle and then the lead vehicle suddenly decelerates. Driver may also be decelerating in traffic lane and then closes in on a decelerating lead vehicle.
- Rear-end—lead vehicle stopped (LVS): driver is going straight and then closes in on a stopped lead vehicle. Driver may also be decelerating or starting in traffic lane and closes in on a stopped lead vehicle. In some of these crashes, the lead vehicle first decelerates to a stop and is then struck by the following vehicle. This typically happens in the presence of a traffic-control device or the lead vehicle is slowing down to turn.
- Negotiating curve—lost control: driver is negotiating a curve and loses control of the vehicle.
- Changing lanes—same direction: driver is changing lanes, passing, or merging and then encroaches into another vehicle traveling in the same direction.
- Turning—same direction: driver is turning left or right at a junction and then cuts across the path of another vehicle initially going straight in the same direction.
- Drifting—same direction: driver is going straight or negotiating a curve and then drifts into an adjacent vehicle traveling in the same direction.
- Road-edge departure—no maneuver: vehicle is going straight or negotiating a curve and then departs the edge of the road at a non-junction area. Vehicle was not making any

maneuver such as passing, parking, turning, changing lanes, merging, or a prior corrective action in response to a previous critical event.
- Opposite direction—no maneuver: vehicle is going straight or negotiating a curve and then drifts and encroaches into the lane of another vehicle traveling in the opposite direction.

Based on crash statistics from the 2004-2008 National Automotive Sampling System General Estimates System (GES) crash databases, light vehicles were involved in crashes preceded by these nine pre-crash scenarios at an average annual frequency of about 2,674,000 police-reported crashes.

Table 1 ranks the target pre-crash scenarios by crash frequency of light vehicle involvement as the subject vehicle. The FCW function deals with rear-end pre-crash scenarios that were associated with 54.7 percent of all target crashes. The LCM function addresses changing lanes and turning pre-crash scenarios that accounted for 18.9 percent of all target crashes. The CSW function addresses the loss of control on a curve crashes that comprised about seven percent of all target crashes. The LDW function addresses the remaining 19.6 percent of all target crashes in which the light vehicle drifted out-of-lane, resulting in road-edge departure, opposite-direction crash, or same-direction crash. The LDW-C function addresses road-edge departure and opposite-direction pre-crash scenarios, whereas the LDW-I function deals with light vehicles involved in the drifting/same direction pre-crash scenario.

Table 1. Annual frequency of target crashes by pre-crash scenario

Pre-Crash Scenario	**Crashes**	**% Crashes**
Rear-end/lead vehicle stopped	907,000	33.9%
Rear-end/lead vehicle decelerating	378,000	14.1%
Road edge departure/no maneuver	371,000	13.9%
Changing lanes/same direction	311,000	11.6%
Turning/same direction	195,000	7.3%
Negotiating a curve/lost control	181,000	6.8%
Rear-end/lead vehicle moving	177,000	6.6%
Opposite direction/no maneuver	103,000	3.9%
Drifting/same direction	51,000	1.9%
Total	2,674,000	100.0%

1.3 Field Operational Test

The field operational test (FOT) included 108 drivers from southeast Michigan, who drove 16 equipped 2006 and 2007 Honda Accords. While an Accord was used as the prototype test vehicle, the research conducted in this field test applies to all light vehicles. The drivers were balanced for gender and age, including younger, middle-aged, and older drivers. These age groups include drivers from ages 20-30, 40-50, and 60-70 respectively. Throughout their participation in the field test, participants drove the instrumented vehicle for any purpose they would use their own personal vehicle for.

The field test started in April 2009 and was completed in early May 2010. The experimental design of the test was an AB design, meaning that each subject experienced two test conditions over a period of 40 days. During the first condition (**A**B), called the baseline period, subjects drove the instrumented vehicle for about 12 days with the integrated safety system turned off. In the second condition (A**B**), or treatment period, subjects drove the vehicle for about 28 days with the integrated safety system enabled. Even though the system alerts were disabled during the baseline period, the on-board data acquisition system (DAS) recorded all data and alerts.

Every test subject completed two survey forms and participated in a debriefing interview. Prior to field test participation, surveys were administered to drivers to collect demographic information and information about their driving history. At the end of their participation in the field test, each driver completed a post-drive survey that contained broad survey items to measure overall attitudes towards the integrated safety system, as well as survey items related to driver acceptance of the system. Most items on the post-drive survey asked drivers to rate various items on a seven-point scale with anchored points ranging from strongly disagree to strongly agree. Survey response types also included yes-no questions, and open-ended questions. Appendix A provides an example of the post-drive survey used. Drivers spent approximately 30-45 minutes completing the survey and then reviewed their answers with a researcher to ensure that all sections had been completed correctly, to clarify responses, and to give drivers an opportunity to discuss any area of interest to them.

1.3.1 Demographics of Field Test Participants

The 108 test participants were evenly balanced by age and gender for a total of 18 drivers in each of the six age/gender groups, as shown in Table 2. In the analysis, the six age and gender groups were considered individually (e.g., younger males) as well as in combination (e.g., all younger drivers).

Table 2. Demographics of field test participants

Age Group	Number of Males	Number of Females	Total
Younger (20-30 years)	18	18	36
Middle-aged (40-50 years)	18	18	36
Older (60-70 years)	18	18	36
Total	54	54	108

The average age of the 36 drivers within each age group is shown in Figure 2. Error bars indicate the minimum and maximum ages within each age group. The overall average age of the 108 drivers was 45 years, the average age of the 54 male drivers was 44 years, and the average of the female participants was 46 years.

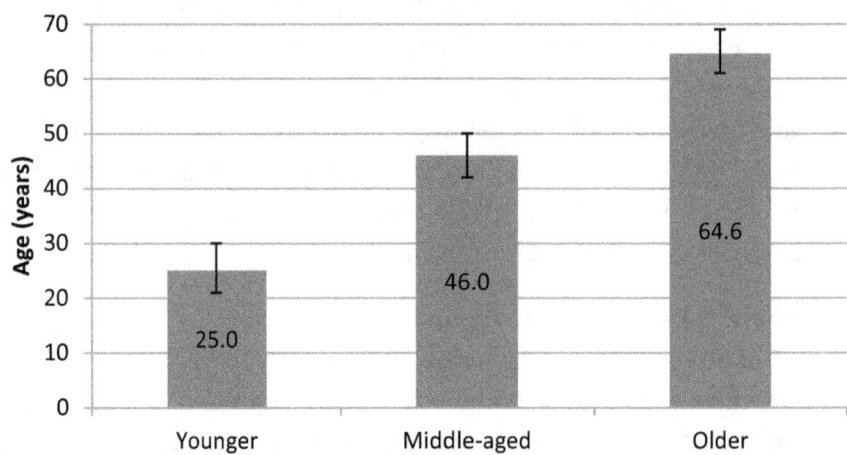

Figure 2. Average age of drivers by age group

The highest level of education attained by each driver is broken down by age group in Figure 3. Sixteen percent of drivers' highest level of education was high school, 42 percent had completed some college, 21 percent held a bachelor's degree, and 19 percent held a master's degree. One driver had a doctorate of philosophy (PhD).

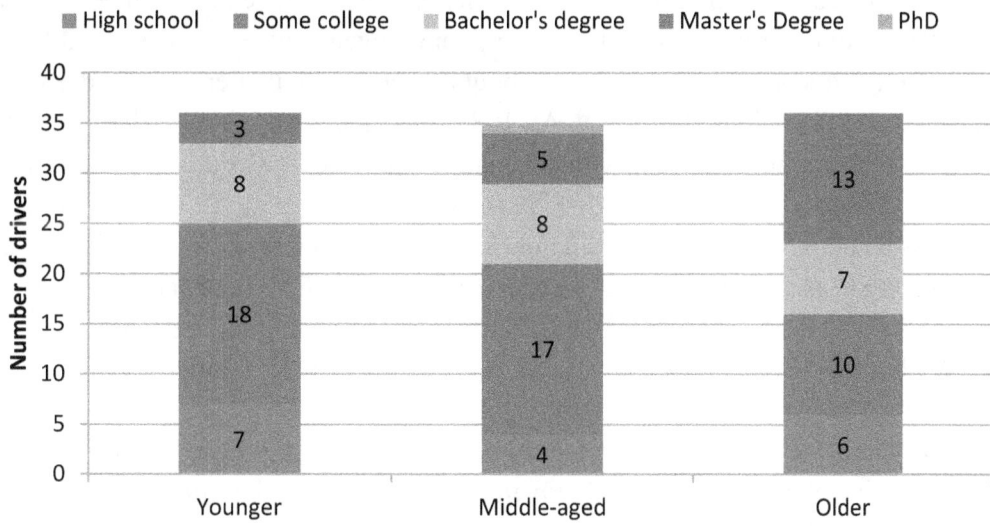

Figure 3. Highest level of education by age group

Overall, drivers had held a driver's license for an average of 28 years with a range of three years to 57 years. Male participants had a driver's license for an average of 27 years, while female participants had held a driver's license for an average of 29 years. Figure 4 shows driving experience broken down by age group, with error bars representing the minimum and maximum number of years of experience in each group.

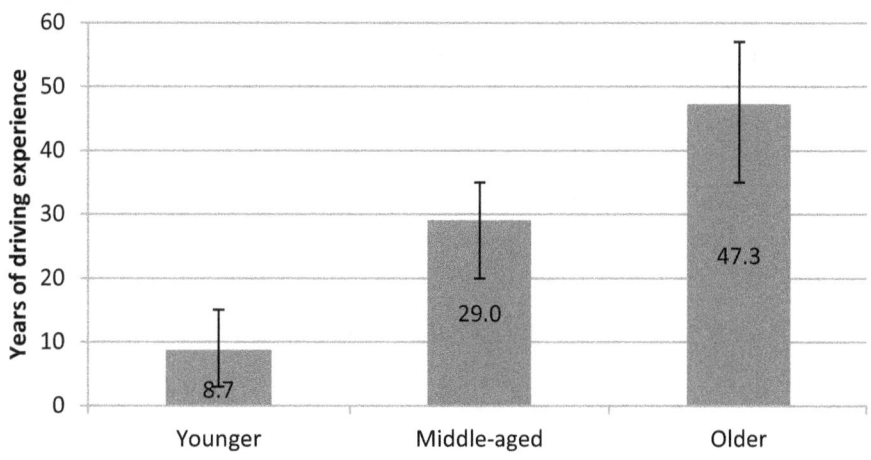

Figure 4. Number of years with a driver's license by age group

1.3.2 Summary of Field Test Exposure

Throughout the course of the field test, drivers accumulated over 213,000 miles of driving; almost 68,898 during the baseline period and 144,496 with the system enabled. Table 3 presents statistics on mileage driven and experience with different system alert types for all drivers in the baseline and treatment periods of the FOT (Sayer et al., 2010). Mileage during the 12-day baseline period ranged from 157 miles to 1,684 miles with an average of 638 miles. Mileage during the 28 days of exposure to the integrated system ranged from 381 miles to 4,307 miles, with an average treatment mileage of 1,338. About 84 percent of all alerts issued during the field test were cautionary drift alerts (LDW-C). All 108 drivers received LDW-C alerts during both the baseline and the treatment periods with an average exposure of 81 during the baseline period (no audible alert issued to the driver) and 78 during the treatment period. The other four alert types were less common. Eight drivers did not experience any FCW alerts, 11 drivers did not experience any CSW alerts, 10 drivers did not experience any LCM alerts, and one driver did not experience an LDW-I alert. During the treatment period, the 108 drivers experienced an average of five FCW alerts, six CSW alerts, eight LCM alerts, and 15 LDW-I alerts.

The number of alerts per 100 miles experienced by drivers in the field test is broken down by treatment period and alert type in Table 4. Overall alert rates in the baseline period ranged from 1.5 to 53.6 alerts per 100 miles, with an average of 14.0 alerts per 100 miles. Alert rates decreased during the treatment period; the driver with the lowest alert rate during the treatment period received 1.7 alerts per 100 miles and the driver with the highest alert rate received 28.8 alerts per 100 miles. The average alert rate across drivers during the treatment period was 8.3 per 100 miles.

Table 3. Exposure of test subjects in the field test

	Baseline						Treatment					
	Miles	FCW	CSW	LCM	LDW-I	LDW-C	Miles	FCW	CSW	LCM	LDW-I	LDW-C
Min	157	0	0	0	0	2	381	0	0	0	0	3
Max	1,684	18	37	22	43	564	4,307	21	58	58	162	446
Average	638	3	3	4	8	81	1,338	5	6	8	15	78

Table 4. Alert rates (number of alerts per 100 miles driven) experienced during the field test

	Baseline						Treatment					
	FCW	CSW	LCM	LDW-I	LDW-C	All	FCW	CSW	LCM	LDW-I	LDW-C	All
Min	0.0	0.0	0.0	0.0	0.7	1.5	0.0	0.0	0.0	0.0	0.3	1.7
Max	2.2	5.2	3.6	6.2	47.8	53.6	1.6	3.3	4.4	6.0	25.7	28.8
Average	0.4	0.4	0.7	1.3	11.2	14.0	0.4	0.4	0.6	1.1	5.8	8.3

Three crashes occurred during the field test. A low speed, rear-end crash occurred in stop-and-go traffic on the freeway. No alert was issued for this crash as it occurred during the baseline period. One driver ran off the road in icy/snowy road conditions but no alert was issued for this crash due to obstructed lane lines. The third crash occurred when a driver sideswiped a construction barrel while exiting the freeway.

1.4 Independent Evaluation

The independent evaluation of the integrated crash warning system had the following goals (Najm et al., 2006):

- *Achieve a detailed understanding of system safety benefits*: estimates the number of crashes that might be avoided by the full deployment of the integrated safety system in light vehicles in the United States. This goal also addresses unintended consequences in terms of modifications in driver behavior that can have negative side effects on traffic safety.
- *Determine driver acceptance of the system*: assesses the ease of use, perceived usefulness, ease of learning, drivers' advocacy, and drivers' assessment of their own driving performance with the integrated safety system.
- *Characterize system performance*: examines the operational performance of the integrated safety system and its components in the driving environment.

1.4.1 Data Processing

Data analysis in the independent evaluation involved many forms of data and data processing procedures. The raw field test data underwent a significant amount of processing in order to synchronize the video with numerical data and to conduct data mining and analysis. Figure 5 presents a flowchart showing each type of data and the data processing procedures. The blocks on the far left of Figure 5 (UMTRI data, video data, and numerical data) represent the raw data received from the University of Michigan Transportation Research Institute (UMTRI), the field test conductor. The blocks at the far right end of the figure (video processing, data mining, data logger, and data viewer) refer to the data types and processes created by the independent evaluator, and the lowest block (data tables) indicates the final output. More detailed information on the data and video processing procedures used to conduct this analysis can be found in Appendix B.

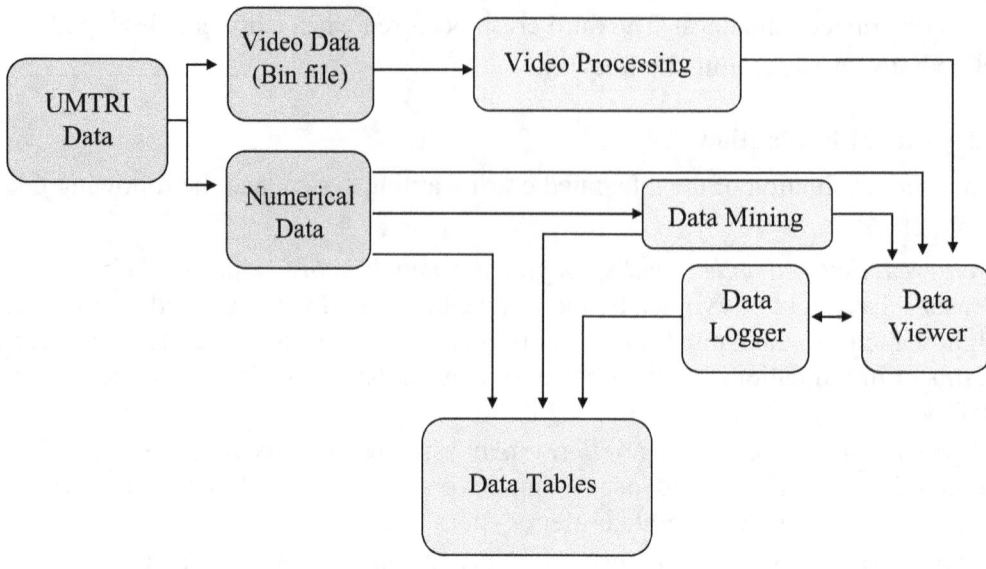

Figure 5. Data processing procedures

The raw video data consist of *.bin* files. Each trip had video files recorded from five different cameras at different sampling rates, as shown in Table 5. The video processing block in Figure 5 represents the process used to convert the raw video files into a format that allows each file to be synchronized with the numerical data and be compatible with the Volpe Center's custom data analysis tool. The first step involved in video processing was to convert the binary video files into standard *.avi* video format. The second step involved was recompression of the *.avi* files to remove any corrupt frames or errors. This conversion and recompression process allows synchronization of the video data at different sampling rates, frame by frame, and with the numerical data by creating a mapping from each numerical data point to the corresponding frame in the video data. This level of synchronization is necessary to extract certain information about system performance.

Table 5. Light vehicle video sampling rates

Camera Type	Sampling Rate
Forward view	10 Hz
Driver's face	10 Hz
Cabin/instrument panel	5 Hz
Left side of truck	5 Hz
Right Side of truck	5 Hz

The raw numerical data was stored in a Structured Query Language (SQL) database format, and consisted of 10 Hz and 100 Hz data. The raw data are processed by data mining algorithms and synchronized with video data so that it can be viewed directly. The data mining block in Figure 5 represents the process under which the data mining algorithms are run on the raw numerical data to produce tables of new variables stored in a separate database.

Once all video had been processed and synchronized with the numerical data, the Volpe Center's data analysis tool was used to extract information about system performance from the videos. This method allowed the analyst access to objective information about the driving scene (e.g., speed, distance to lead vehicle, turn signal usage) as a supplement to the video. As the video is reviewed, objective information is extracted and entered into the data logger and then stored in a numerical database. The results of the data mining algorithms and video analysis, as well as some of the raw numerical data, are then extracted using SQL queries. The data tables, shown at the bottom of Figure 5, were used to conduct all analyses.

1.4.2 Multimedia Data Analysis Tool

The Volpe Center developed a multimedia data analysis tool (MDAT) to extract objective information from the five video data channels collected during the field test. While the numerical data provide information about vehicle dynamics and the driving scenario, some information can only be obtained from examining the video. Video analysis is used to supplement the numerical data.

The MDAT is used to synchronize and simultaneously play back five video channels, presenting a full view of the driving scene and driver. In addition to video data, the MDAT is connected directly to the numerical database and can display any of approximately 200 numerical data channels along with the video. Synchronizing video with numerical data allows the viewer full access to all of the information necessary to fully assess the driving scenario and driver condition.

Figure 6 shows a screen view of the MDAT. The left side of the viewing window shows five channels of video data: front road scene, driver face, cabin, left-side road scene, and right-side road scene. The video is controlled by the buttons on the bottom of the window and the numerical data could be displayed in a separate window. Drop-down menus on the right side of the screen are provided to code specific information about the video as viewers watch videos of interest. The information entered in these menus is saved in a table as part of the field test database, making it accessible to support further analysis. In this analysis, a total of 16,915 light vehicle alerts were viewed and coded. This analysis included all imminent alerts for each driver (FCW, CSW, LCM and LDW-I), and a random sample of about 65 percent of cautionary lane-departure warnings (LDW-C). The latter warnings were sampled because they accounted for 75 percent of all alerts issued in the field test. A breakdown of the video analysis by alert type and treatment period is shown in Table 6. Detailed information about the video sampling and video analysis can be found in Appendix C and Appendix D.

Figure 6. Screen view of multimedia data analysis tool

Table 6. Number of alerts included in video analysis by alert type

Alert Type	Baseline	Treatment	Total
FCW	274	567	841
CSW	311	587	898
LCM	413	907	1,320
LDW-I	874	1,584	2,458
LDW-C	5,483	5,662	11,145
All	7,355	9,307	16,662

2. Safety Impact

This analysis addresses the safety benefits goal of the independent evaluation by asking two key questions:
- If all light vehicles in the United States were equipped with the integrated safety system, what would be the annual change in the total number of rear-end, lane-change, and road-departure crashes?
- Would use of the integrated safety system result in unintended consequences that might impact overall traffic safety in a negative or positive manner?

The first question deals with the estimation of potential safety benefits that would result from full deployment of integrated safety systems. The second question looks for any unintended driving behavior from system use that could potentially cause harm to the equipped vehicle or other road users.

The integrated safety system was designed as a countermeasure to a number of pre-crash scenarios that occur immediately prior to rear-end, lane-change, and road-departure crashes (Najm et al., 2007). The safety benefits are derived from system effectiveness in reducing the frequency of target pre-crash scenarios listed in Table 1. The LDW function may also prevent opposite-direction crashes due to unintentional drifting into a left-adjacent lane of oncoming traffic.

> **HIGHLIGHTS**
> - Based on reduction rates of near crashes, about 475,000±313,000 police-reported crashes could be prevented annually if all light vehicles in the United States were equipped with the integrated safety system.
> - No negative unintended consequences were observed with the integrated safety system.
> - Drivers showed a 21 percent increase in turn signal usage when driving with the system.
> - Driving with the system resulted in an overall 21 percent reduction in the rate of lane departures.
> - Half of the drivers who attended a focus group session reported that the system prevented them from getting into a crash.
> - For speeds over 55 mph, there was an overall decrease in driving conflict rates.
> - All drivers showed a reduction in lane-change and road departure near crashes with the system enabled; the rate of lane-change near crashes decreased for men and the rate of road departure crashes decreased more for women.
> - Near crashes decreased by 19 percent in younger drivers.

2.1 Safety Impact Technical Approach

Figure 7 illustrates the analysis framework used to assess safety impact. This framework divides the test subjects' driving experience into three areas: overall experience; driving conflicts; and near crashes. In this research, overall driving includes all data collected during the field test. Driving conflicts refer to a small subset of the overall driving data that in which, had the driver not intervened (by steering or using the brake) it is likely that a crash would have occurred. Near crashes are made up of a small subset of the driving conflict scenarios: those with the most extreme driver reactions. The safety analysis compares the test subjects' driving experience in each area between the baseline (B) and treatment periods.

In addition to comparing driver behavior between the baseline and treatment conditions, this analysis looked into longer-term adaptation to the system by comparing performance in the baseline condition to the second half of the treatment condition, or T_2. During the first part of the treatment period (T_1), it is possible that drivers' behavior changed due to the presence of a new system in their vehicle. The intent of analyzing T_2 separately was to analyze only data collected after drivers had gone through an initial adaptation period. It should be noted that the duration of the baseline condition was about 12 days, while participants drove in the treatment condition for approximately 28 days. The treatment condition was split into two treatment periods by miles driven, each accounting for approximately half of the miles driven during the treatment condition. The mileage of the treatment periods varies slightly because individual trips were not divided and assigned to two different test periods. Throughout this report, treatment, or T_{all}, represents data from the entire 28 day treatment period, and T_2 represents data from the second half of each driver's treatment exposure.

Results from the analysis were synthesized to project potential safety benefits. Safety benefits are expressed in terms of the system's potential to reduce the number of target crashes. These benefits are ideally measured from actual crash data; however, only three crashes were observed during the field test. Thus, this analysis estimates the safety benefits by applying a methodology that uses non-crash performance data (driver, vehicle and system) collected during the field operational test (Ference et al., 2006).

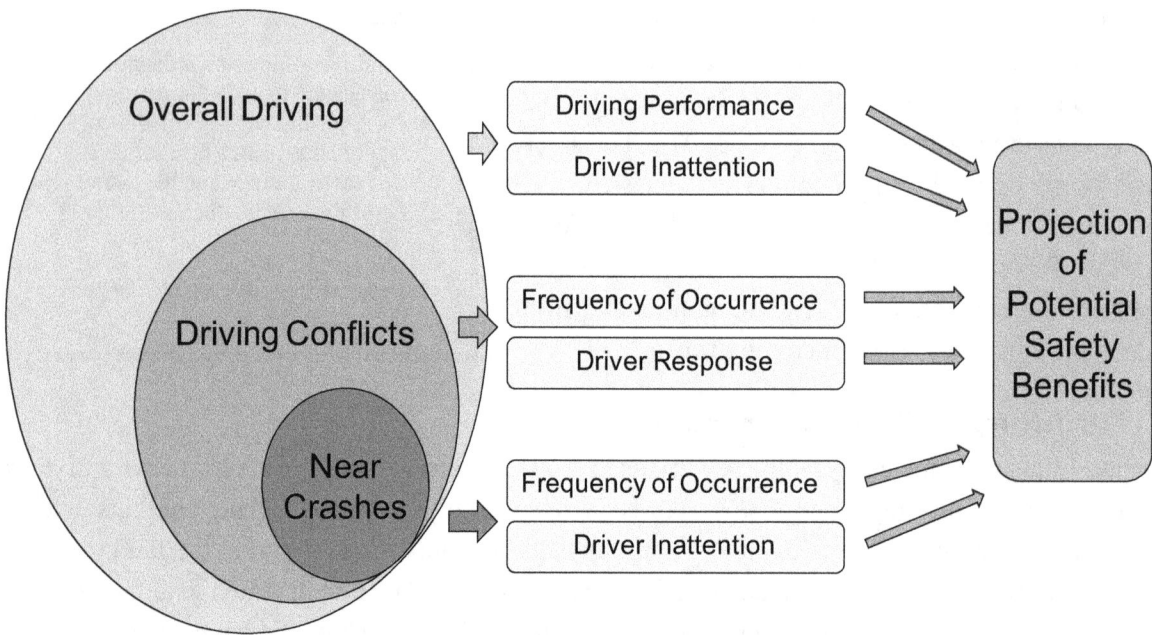

Figure 7. Safety benefits framework

2.2 Overall Driving Experience

This analysis addresses driver performance in the overall driving experience captured during the field test. This analysis was undertaken to examine any unintended consequences resulting from the use of the integrated safety system. Driving measures were compared within subjects in each age/gender category.

Two-tail paired *t*-tests were performed to compare driver performance. A paired *t*-test is used to determine if there is a statistically significant difference between the means of the same subjects under different circumstances. A two-tailed test is used when the mean under the test condition can be either greater than or less than the mean during baseline. For all *t*-tests conducted in this analysis, a *p* value of 0.05, or 95 percent confidence, was used to define statistical significance. These values are indicated by bold font in the tables throughout this report.

The measures used in this analysis for driving performance include the following:
- Speed maintenance (travel speed)
- Time headway
- Lane changes:
    - Number of lane changes per 100 vehicle miles driven
    - Proportion of signaled lane-change maneuvers
- Lane keeping:
    - Number of lane busts[1] per 100 miles driven
    - Mean duration of lane bust events

Each of the above measures was broken down by weather condition (clear/adverse), lighting (light/dark) and road type (freeway/non-freeway).

The measures for inattentive behavior include:
- Secondary tasks: the proportion of analyzed alerts with secondary tasks
- Eyes-off-forward-scene: the proportion of analyzed alerts with eyes-off-forward-scene

The analysis of the overall driving experience was conducted in two period comparisons using paired *t*-test for means between:
- Baseline and T_{all} : the entire treatment period
- Baseline and T_2: the second half of the treatment period

The analysis was also broken down by gender and by three age groups: younger, middle-aged, and older.

---

[1] Lane busts refer to a scenario where any of the vehicle's wheels cross the lane line of the lane the vehicle is currently traveling in when the turn signal is not activated.

2.2.1 Speed Maintenance

The average speed of each driver was calculated for all periods in which the vehicle speed was greater than 35 mph (15.6 m/s). Very little change in average speed was observed, with no statistically significant change overall. Table 7 presents the results of the paired *t*-tests for this dataset.

Table 7. Results of baseline versus treatment paired *t*-test for average speed in m/s

	Overall	Gender		Age (years)		
		Male	Female	20-30	40-50	60-70
Baseline vs. T_{all}:						
B	24.7	25.3	24.2	24.7	25.6	23.9
T_{all}	24.9	25.4	24.3	25.1	25.5	23.9
p	0.55	0.68	0.67	0.18	0.69	0.97
n	108	54	54	36	36	36
Baseline vs. T_2:						
B	24.7	25.3	24.2	24.7	25.6	23.9
T_2	24.9	25.4	24.5	25.2	25.5	24.0
p	0.39	0.825	0.34	0.19	0.72	0.78
n	108	54	54	36	36	36

2.2.2 Headway Keeping

Headway describes how closely the subject vehicle follows a lead vehicle. The measurement is expressed in seconds, and is defined as the time in which the subject vehicle, at its current speed, would reach the current position of the lead vehicle. For this analysis, average headways were calculated over all following events lasting longer than one second, for which the subject vehicle was traveling greater than 25 mph (11.2 m/s).

Results show a small overall decrease in following headways, meaning drivers tended to follow more closely in the treatment period than in the baseline as shown in Table 8. Statistically significant decreases were observed for middle-aged drivers, non-freeway driving, and daytime adverse weather condition. The largest difference was in the daytime adverse weather condition, with headways approximately 0.24 second shorter (a 12% decrease) with the system enabled. It should be noted that headways were generally lower during T_2 than during T_{all}.

2.2.3 Lane-Change Behavior

Each driver's lane-changing behavior was analyzed using two separate measurements, both taken only for periods when vehicle speed was greater than 45 mph (20.1 m/s).

The first measure, lane change rate, refers to the overall number of lane changes per 100 miles driven. The results of the paired *t*-tests for this dataset are given in Table 9. No significant increase or decrease was observed in lane change rate for any category, or overall.

Table 8. Results of baseline versus treatment paired *t*-test for mean headway in seconds

	Overall	Gender		Age (years)			Road Type	
		Male	Female	20-30	40-50	60-70	Freeway	Non-Freeway
Baseline vs. T_{all}:								
B	1.73	1.72	1.74	1.58	1.65	1.97	1.41	2.05
T_{all}	1.69	1.68	1.71	1.54	1.60	1.94	1.37	1.98
p	0.06	0.12	0.26	0.37	**0.05**	0.46	0.16	**0.00**
n	108	54	54	36	36	36	108	108
Baseline vs. T_2:								
B	1.73	1.72	1.74	1.58	1.65	1.97	1.41	1.73
T_2	1.66	1.65	1.66	1.50	1.58	1.89	1.34	1.96
p	**0.00**	**0.02**	**0.04**	0.12	**0.05**	**0.05**	**0.02**	**0.00**
n	108	54	54	36	36	36	108	108

Table 9. Results for overall number of lane changes per 100 miles driven

	Overall	Gender		Age (years)			Road Type	
		Male	Female	20-30	40-50	60-70	Freeway	Non-Freeway
Baseline vs. T_{all}:								
B	43.5	41.0	46.1	50.5	42.1	38.0	49.1	38.2
T_{all}	43.2	41.4	45.0	50.4	43.1	36.0	48.8	36.5
p	0.68	0.75	0.36	0.96	0.42	0.17	0.87	0.35
n	108	54	54	36	36	36	107	108
Baseline vs. T_2:								
B	43.5	41.0	46.1	50.5	42.1	38.0	49.1	38.2
T_2	43.1	41.7	44.5	49.5	43.5	36.3	48.7	36.4
p	0.65	0.57	0.28	0.64	0.28	0.28	0.77	0.35
n	108	54	54	36	36	36	107	108

The second measure is the signal ratio, which is the proportion of lane changes for which the turn signal was used. Results from the analysis of signal ratio in Table 10 show a striking increase in the proportion of signaled lane changes, across both genders, all ages, and both freeway and non-freeway road types. When the integrated system was enabled, drivers received a cautionary drift warning if they crossed the lane line without using their turn signal. This warning encourages the use of turn signal usage because drivers are likely to use their turn signal to avoid getting a warning.

Drivers used their turn signal during an average of 62 percent of lane changes during the baseline period. During the treatment period, turn signal usage increased to 75 percent of lane changes. The largest increases in turn signal usage were seen in males and in middle-aged drivers, both of which showed an increase of 16 percentage points with the system enabled. Both of these groups however, had lower-than-average turn signal usage during the baseline period.

Table 10. Results for ratio of signaled lane changes to total lane changes

	Overall	Gender		Age (years)			Road Type	
		Male	Female	20-30	40-50	60-70	Freeway	Non-Freeway
Baseline vs. T_{all}:								
B	0.62	0.56	0.69	0.67	0.60	0.61	0.65	0.56
T_{all}	0.75	0.72	0.78	0.78	0.76	0.72	0.78	0.68
p	**0.00**	**0.00**	**0.00**	**0.00**	**0.00**	**0.00**	**0.00**	**0.00**
n	108	54	54	36	36	36	107	107
Baseline vs. T_2:								
B	0.62	0.56	0.69	0.67	0.60	0.61	0.65	0.56
T_2	0.75	0.71	0.78	0.78	0.76	0.70	0.78	0.68
p	**0.00**	**0.00**	**0.00**	**0.00**	**0.00**	**0.02**	**0.00**	**0.00**
n	108	54	54	36	36	36	107	106

2.2.4 Lane Keeping

Lane keeping is quantified in terms of "lane busts" – partial or incomplete lane changes in which the host vehicle crossed a lane boundary but returned to its original lane. The measures applied to each driver were number of busts per 100 miles driven and mean duration of busts. The measures included both freeway and non-freeway driving for periods in which the vehicle speed was constant and greater than 25 mph. Events shorter than one second were excluded from the analysis.

A marked decrease was observed in the rate of lane busts across all drivers in both treatment conditions, indicating that drivers maintain better lane positioning when the system is enabled. Drivers experienced an overall 21 percent decrease in the rate of lane busts during the treatment period. The largest proportional decrease was 26 percent for middle-aged drivers. The largest absolute decreases were observed for low speeds (less than 45 mph) and for non-freeway driving, two subsets which largely overlap. Baseline lane bust rates were much higher than average for these categories however, so the proportional changes were more moderate.

Table 11 provides the means and *p* values of the paired *t*-tests associated with lane busts (greater than 0.1 m) per 100 miles traveled in these two speed ranges. Results broken down by speed bin (a discreet value that encompasses a range of speeds) can be found in Table 36 in Appendix E.

Table 11. Results of baseline versus treatment paired *t*-test for lane busts per 100 miles driven

	Overall	Gender		Age (years)			Road Type	
		Male	Female	20-30	40-50	60-70	Freeway	Non-Freeway
Baseline vs. T_{all}:								
B	38.70	37.10	40.30	41.10	40.40	34.50	20.60	55.90
T_{all}	30.60	29.20	32.00	33.23	29.60	28.99	15.40	44.90
p	0.00	0.00	0.00	0.00	0.00	0.01	0.00	0.00
n	108	54	54	36	36	36	108	108
Baseline vs. T_2:								
B	38.70	37.10	40.30	41.10	40.40	34.50	20.60	55.90
T_2	31.10	30.18	32.00	34.09	29.50	29.72	16.00	45.50
p	0.00	0.00	0.00	0.01	0.00	0.02	0.00	0.00
n	108	54	54	36	36	36	108	108

The system's impact on the duration of lane bust events is less clear. Overall, the average duration of lane busts showed a statistically significant decrease by 0.08 second, a three-percent change. However, the change for period T_2 was not significant. These results are shown in Table 12.

Only middle-aged drivers showed a significant decrease in lane bust duration for both treatment periods. The decrease in overall treatment was 0.15 second, a six percent change. A statistically significant decrease of 0.12 second (four percent) was observed for the speed bracket of 35-45 mph. Lane bust duration results by speed bin are located in Table 37 in Appendix E.

Table 12. Results of paired *t*-test for lane bust duration in seconds

	Overall	Gender		Age (years)			Road Type	
		Male	Female	20-30	40-50	60-70	Freeway	Non-Freeway
Baseline vs. T_{all}:								
B	2.72	2.70	2.74	2.72	2.72	2.72	2.51	2.81
T_{all}	2.64	2.59	2.69	2.70	2.56	2.65	2.46	2.76
p	**0.03**	**0.02**	0.38	0.61	**0.01**	0.44	0.43	0.21
n	108	54	54	36	36	36	108	108
Baseline vs. T_2:								
B	2.72	2.70	2.74	2.72	2.72	2.72	2.51	2.81
T_2	2.66	2.64	2.68	2.75	2.56	2.68	2.46	2.80
P	0.17	0.32	0.35	0.66	**0.03**	0.63	0.54	0.97
N	108	54	54	36	36	36	108	108

2.2.5 Vehicle Speed at Curve Entry

Three different measures were used to assess changes in how drivers approached curves: vehicle speed three seconds prior to curve start, vehicle speed five seconds prior to curve start, and average acceleration over the five seconds prior to curve start. Each measure was taken only for curves of travel duration greater than three seconds, with radius between 100 and 1000 meters.

None of the three measures showed any significant change, even when broken into smaller groups based on driver age and gender, road type, and curve radius (Table 38 in Appendix E).

2.2.6 Attention to Primary Driving Task

This analysis focused on driver attention to the driving task and the forward scene for all alerts analyzed in the video analysis. Secondary tasks and eyes-off-forward scene events were recorded for each alert, as discussed in Section 2.2. See Appendix D for the list of secondary tasks and definition of eyes-off-forward scene. The analysis was broken down by age and gender. For each analysis, data were compared between the baseline and treatment periods.

2.2.6.1 Analysis of Secondary Tasks

Secondary tasks include behaviors exhibited by the driver that do not support the primary driving task and could be potentially distracting to the driver. These tasks were extracted by viewing the face and cabin cameras of the 16,915 analyzed videos. Figure 8 lists the 10 secondary tasks that drivers engaged in most frequently during the field test, and the percentage of alerts analyzed in which each task was observed. The most frequent secondary task was communicating with or looking at passengers in the vehicle, followed by grooming (scratching face, rubbing eyes, combing hair, etc.), and talking on a cellular phone. Overall, cell phone related activities were present in 13 percent of the alerts analyzed. Eight of the 10 most common secondary tasks observed in this field test are consistent with the secondary tasks observed in a field test conducted by the Virginia Tech Transportation Institute (Neale et al., 2005). The two tasks that differed, reading cell phone and text messaging, were not coded individually by Virginia Tech.

The driver who exhibited the fewest secondary tasks, a middle-aged female, engaged in secondary tasks in only 17 percent of alerts, eight percent of which involved talking on the phone. The driver who engaged in secondary tasks the most frequently, a younger female, performed secondary tasks in 87 percent of the analyzed episodes. This driver was observed talking on a cellular phone in 27 percent of alerts, grooming in 26 percent of alerts, talking to passengers in 17 percent of alerts, and text messaging in 14 percent of alerts.

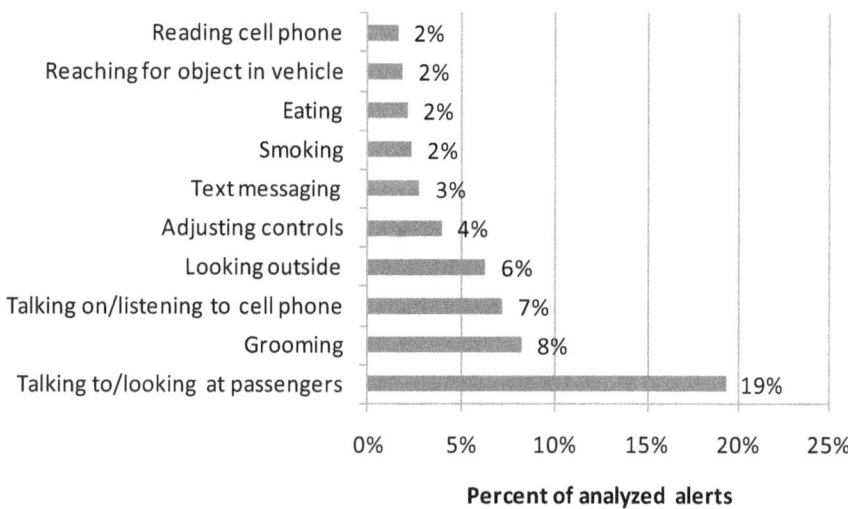

Figure 8. Most frequent secondary tasks exhibited during the field test

The types of secondary tasks drivers engaged in varied by age group. Figure 9 lists the four most frequent secondary tasks for each age group. While the most frequent secondary task for all three age groups was talking with passengers, middle-aged drivers engaged in this behavior less frequently than the older and younger drivers. Younger drivers were more likely than middle-aged and older drivers to engage in cell phone related behavior; younger drivers were talking on a cell phone in 10 percent of the alerts analyzed compared to seven percent and five percent for middle-aged and older drivers. The most frequent observance of talking on a cellular phone was by a younger female, who was seen talking on the phone in 39 percent of her alerts. Nineteen of the drivers were never observed talking on their cell phone while driving. Overall, younger drivers were observed text messaging in five percent of the videos analyzed, while the other two age groups were observed text messaging in less than three percent of the videos. One driver, a younger female, was observed to be text messaging in 19 percent of her alert episodes. Just under half of the drivers (50) were observed text messaging.

The most frequent use of a Bluetooth device was by a middle-aged male, who was talking on his Bluetooth in 29 percent of episodes, and only six of the drivers regularly used a Bluetooth headset. Eighteen of the drivers were smokers. The most infrequent behavior observed in the field test, "eyes closed for greater than one second" was exhibited by only eight of the drivers. Seven of the eight drivers exhibited this behavior only one time, and the eighth (a middle-aged male) was only observed exhibiting this behavior two times.

The proportion of each driver's alerts in which they were engaged in any secondary tasks was broken down by treatment period to observe the changes in drivers' secondary-task engagement over time. As shown in Figure 10, younger males and older females engaged in slightly more secondary tasks during T_1, but these habits were not sustained during T_2. Overall, younger females showed the most frequent secondary-task engagement, and older females showed the least. The data show no overall trend in changes in secondary-task engagement with the integrated system enabled.

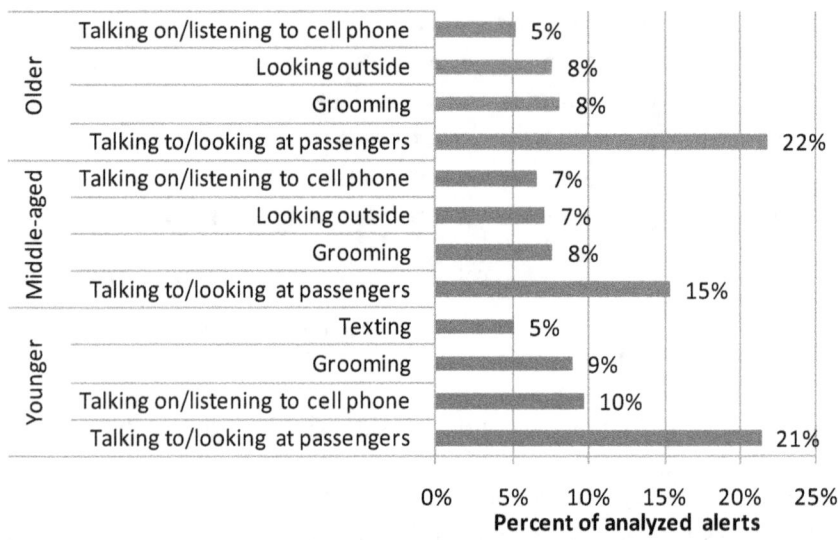

Figure 9. Most frequent secondary tasks exhibited during the field test, by age group

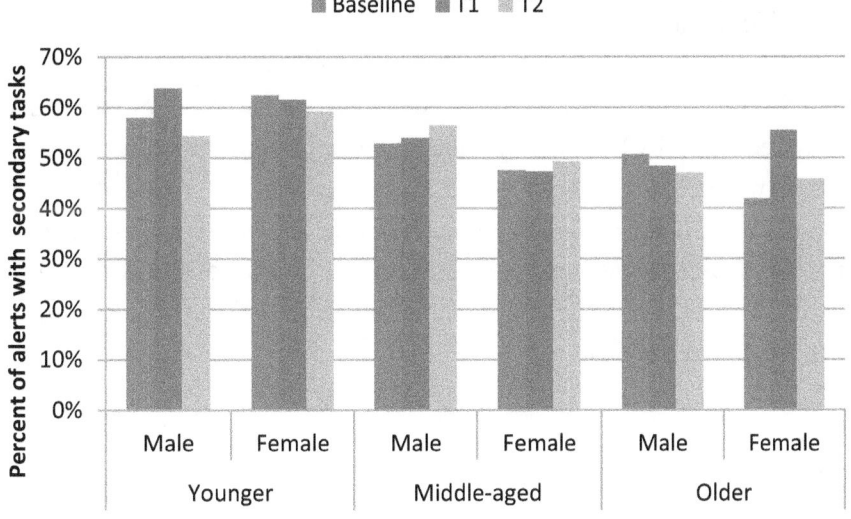

Figure 10. Percent of alerts with secondary tasks by age, gender group, and treatment period

Table 13 shows the results of paired *t*-tests that compared drivers' secondary-task engagement between test periods. None of the age and gender groups showed a significant increase or decrease in secondary-task engagement with the system enabled.

Table 13. Paired *t*-test for percent of analyzed alerts with secondary tasks

	Overall	Gender		Age (years)		
		Male	Female	20-30	40-50	60-70
Baseline vs. T_{all}:						
B	6.6%	7.6%	5.6%	7.5%	6.8%	5.6%
T_{all}	6.0%	7.0%	5.1%	7.5%	6.0%	4.6%
P	0.34	0.51	0.48	1.00	0.36	0.28
N	107	54	53	36	35	36
Baseline vs. T_2:						
B	6.6%	7.6%	5.6%	7.5%	6.8%	5.6%
T_2	6.7%	7.7%	5.8%	7.5%	7.1%	5.6%
P	0.86	0.94	0.86	0.95	0.79	1.00
N	107	54	53	36	35	36

2.2.6.2 *Analysis of Eyes-Off-Forward-Scene*

In the video analysis of the alert episodes, "eyes-off-forward-scene" was defined as an instance when the driver had his eyes diverted from the forward driving scene for at least 1.5 continuous seconds in the five seconds leading up to the alert. Eyes-off-forward-scene behavior was observed in seven percent of the episodes analyzed. The driver with the highest proportion of eyes-off-forward-scene behavior during the field test was a younger male, who had his eyes off the forward scene in 23 percent of the alerts analyzed. Twelve of the drivers showed no instances of eyes-off-forward-scene.

To illustrate how drivers' eyes-off-forward-scene behavior changed over the course of the field test, data were broken down by treatment period. Figure 11 shows the proportion of alerts in which drivers' eyes were off the forward scene by age and gender group. All six groups showed a slight decrease in eyes-off-forward-scene behavior when the system was first enabled (T_1), but for most groups, this decrease was not sustained during T_2. The results of the paired *t*-test comparing eyes-off-forward-scene behavior, shown in Table 14, show no statistically significant changes between treatment periods.

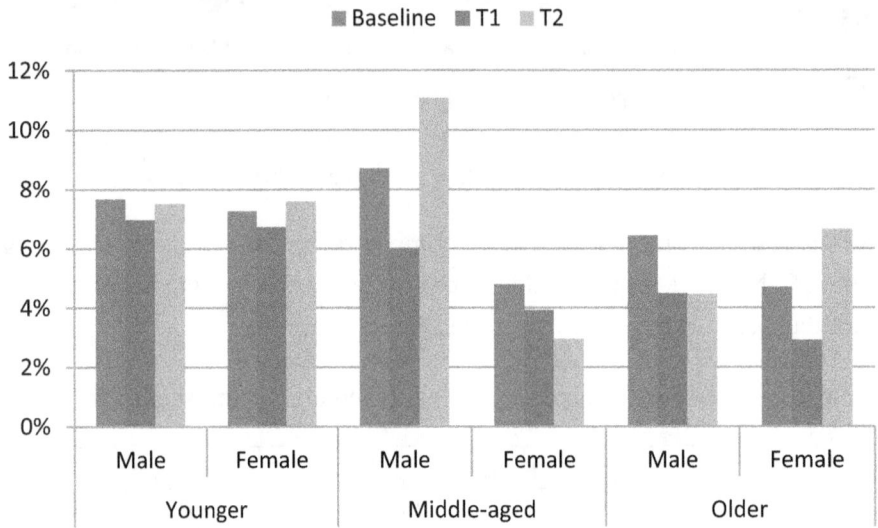

Figure 11. Percent of eyes off forward scene by treatment period by age and gender group

Table 14. Results of paired *t*-test for percent of analyzed alerts with eyes off forward scene

	Overall	Gender		Age (years)		
		Male	Female	20-30	40-50	60-70
Baseline vs. T_{all}:						
B	48%	46%	49%	40%	50%	54%
T_{all}	46%	46%	47%	40%	48%	51%
p	0.28	0.83	0.20	0.89	0.41	0.27
n	107	54	53	36	35	36
Baseline vs. T_2:						
B	45%	46%	49%	40%	50%	54%
T_2	48%	47%	49%	43%	47%	54%
p	0.85	0.55	0.74	0.16	0.34	0.97
n	107	54	53	36	35	36

2.3 Conflict Exposure Rates and Driver Response

Analyses were conducted to determine whether there was a change from the baseline to the treatment period in conflict exposure rates and/or in the drivers' response to conflicts. These analyses were conducted for all conflict types combined as well as separately for the following types of conflicts:

- Rear-end driving conflicts:
    - Lead vehicle decelerating (LVD)
    - Lead vehicle moving at slower constant speed (LVM)
    - Lead vehicle stopped (LVS) values were not analyzed because of a high error rate in the performance of the sensors for that alert type
- Lane-change conflicts:
    - Lane changes to the right (LCR)
    - Lane changes to the left (LCL)
    - Turning conflicts could not be analyzed due to an insufficient number of occurrences
- Road-departure conflicts:
    - Departing straight road to the left (SDL) and right (SDR)
    - Departing curved road to the left (CDL) and right (CDR)
- Approaching a curve with excessive speed (CES)

Specific thresholds used to determine conflict scenarios can be found in Appendix F.

It is possible that the integrated system had an effect only under specific conditions and that this effect was undetectable in the overall analysis (all drivers under all conditions). Consequently, for each conflict type additional tests were run using only the following subsets of the data:

- Age group (20-30, 40-50 and 60-70 year olds)
- Gender (male and female)
- Speed (25-55 mph and 55 mph and over)
- Lighting (day and night) *(only conducted for conflict rate, not driver response)*
- Weather (clear and adverse) *(only conducted for conflict rate, not driver response)*

Conflict rate was defined as the number of conflicts experienced per 100 miles driven. Drivers who drove less than 100 miles in a given condition were excluded from the analysis of that condition. Driver response was defined using different measures depending on the type of conflict being analyzed. Those measures are defined in Table 15. By definition, a driving conflict scenario requires that the driver take evasive action to avoid a potential collision.

Table 15. Measures used to quantify driver response

Conflict Type	Measures Analyzed	Units
Rear End	Time-to-collision at brake onset	s
	Minimum time-to-collision during conflict resolution	s
	Peak deceleration level during conflict resolution	m/s^2
	Average deceleration level during conflict resolution	m/s^2
	Headway time at brake onset (only for LVD)	s
Lane Change, Road Departure	Maximum lateral acceleration (straight roads only)	m/s^2
	Average lane bust time	s
	Maximum lane bust distance	m
Curve Speed	Maximum lateral acceleration	m/s^2
	Change in speed from start of curve to tightest point in curve	m/s

The results of the analyses of conflict rate, driver response and LDW impact are presented in the following sections and, where indicated, in the appendices. The sections are organized in terms of conflict type, presenting both conflict rate and driver response together for each. In the tables, where dashes appear they represent comparisons for which there were too few subjects for statistical testing ($n < 8$).

2.3.1 Overall Driving Conflict Rate

There was no statistically significant change in overall conflict rates from baseline to the treatment period (Table 39 in Appendix G). Likewise, total conflict rate did not vary within any of the subgroups, such as males, females, age groups, light or weather conditions—except for speeds above 55 mph: from baseline to T_{all} the total conflict rate dropped from 4.14 to 3.71 conflicts per 100 miles driven ($p = 0.03$, $n = 99$), and to T_2 it dropped from 4.14 to 3.69 conflicts per 100 miles driven ($p = 0.07$, $n = 99$). However, the lack of an overall effect can be misleading since significant effects in one type of driving conflict may be masked by a lack of an effect, or an effect in the opposite direction, in another driving conflict type. Consequently, the different conflict types are analyzed individually in the following sections.

2.3.2 Rear-End Driving Conflicts

Conflict Rate. There was no significant change in the overall rates at which drivers experienced rear-end conflicts, either with vehicles that were decelerating (LVD) or moving at a constant speed (LVM) (Table 40 in Appendix G). The only statistically significant effect was for LVM: when only nighttime driving was considered, the conflict rate decreased from 1.18 to 0.78 conflicts per 100 miles driven from baseline to T_{all} ($p = 0.03$, $n = 29$). However, from baseline to T_2 the decrease (1.23 to 0.91 conflicts per 100 miles driven) was not statistically significant ($p = 0.13$, $n = 25$).

For the following measures of driver response to conflicts, the results are presented in Table 41 in Appendix G. For these analyses, only responses involving braking were used; those involving only steering or the gas pedal were omitted.

Time-to-Collision at Brake Onset. An advanced warning system might be expected to cause an increase in the minimum time-to-collision at brake onset, since drivers would be able to hit the brakes earlier. However, average time-to-collision values at brake onset did not consistently increase or decrease for either LVD or LVM, and none of the changes were significant. For older drivers in LVM, there was a trend towards a longer time-to-collision in T_{all} (3.38 s) as compared to baseline (3.17 s) ($p = 0.07$, $n = 27$), but no such trend existed between T_2 (3.36 s) and baseline (3.29 s) ($p = 0.69$, $n = 25$).

Minimum Time-to-Collision during Conflict Resolution. The minimum time-to-collision represents the intensity of a conflict by indicating how close a driver was to hitting the car in front. Again, average overall times did not change significantly. There was, however, for women a significant decrease from baseline to T_2 in the minimum time-to-collision for LVM: 2.54 to 2.43 seconds ($p = 0.04$, $n = 33$). In other words, these women were closer to potentially colliding with a slower-moving vehicle with the system enabled.

Peak Deceleration Level during Conflict Resolution. The peak deceleration indicates the response intensity. For both decelerating and constant-speed lead vehicles, the minimum deceleration did not vary significantly in either test period, overall or for the various subcategories such as gender, age or speed.

Average Deceleration during Conflict Resolution. There was a trend for women to brake harder in the treatment group (-1.63 m/s^2) than in the baseline (-1.47 m/s^2) in LVM ($p = 0.07$, $n = 38$), but there was no such trend or significant result for baseline compared to T_2, or for any of the other groups or for the overall average deceleration.

Headway Time at Brake Onset. There was no statistically significant change in headway time at brake onset, either overall or for any of the individual categories.

2.3.3 Lane-Change Driving Conflicts

For all of the following analyses, both for conflicts and driver responses, events occurring while the vehicle was moving under 25 mph were omitted since the lane tracking function was not operable at low speeds.

Conflict Rate. There was an overall significant decrease in lane-change conflicts to the left for both treatment periods as seen in Table 16. This overall decrease was not shown for lane-change conflicts to the right; however, there was a significant drop in lane-change conflicts to the right at slower speeds (25-55 mph) and during the day from baseline to T_{all}.

Maximum Lateral Acceleration. Maximum lateral acceleration indicates the strength of response to a lane-change conflict. It was not possible to analyze driver response in terms of lateral acceleration on curved roads because precise knowledge of the acceleration necessary to make the curve was not known and thus could not be subtracted to get the acceleration of the road-departure correction. Consequently, the analysis could only be conducted for straight roads.

Overall the system had no significant effect on maximum lateral acceleration, except in the case of lane-change conflicts to the left among middle-aged drivers: the average lateral acceleration increased from 0.75 to 0.86 m/s^2 from baseline to T_{all} ($p = 0.05$, $n = 21$), and from 0.74 to 0.89 m/s^2 from baseline to T_2 ($p = 0.02$, $n = 17$) (Table 42 in Appendix G). This is different from the younger and older age groups, both of which decreased their acceleration, although not statistically significantly.

Table 16. Average number of lane-change conflicts per 100 miles driven

Lane change to the right (LCR):

	Overall	Gender		Age (years)			Light		Weather		Speed (mph)	
		Male	Female	20-30	40-50	60-70	Night	Day	Clear	Adverse	25-55	55+
Baseline vs. T_{all}:												
B	0.45	0.41	0.50	0.47	0.43	0.45	0.51	0.64	0.47	-	0.52	0.87
T_{all}	0.39	0.35	0.44	0.43	0.38	0.37	0.51	0.46	0.41	-	0.22	0.80
p	0.06	0.14	0.25	0.37	0.36	0.18	0.96	**0.01**	0.11	-	**0.02**	0.31
n	53	29	24	17	17	19	8	46	52	-	9	45
Baseline vs. T_2:												
B	0.50	0.45	0.56	0.56	0.47	0.49	-	0.74	0.52	-	-	0.88
T_2	0.47	0.41	0.52	0.54	0.37	0.50	-	0.57	0.48	-	-	0.86
p	0.47	0.49	0.68	0.84	0.06	0.93	-	0.11	0.57	-	-	0.79
n	40	20	20	12	14	14	-	32	39	-	-	35

Lane change to the left (LCL):

	Overall	Gender		Age (years)			Light		Weather		Speed (mph)	
		Male	Female	20-30	40-50	60-70	Night	Day	Clear	Adverse	25-55	55+
Baseline vs. T_{all}:												
B	0.60	0.41	0.86	0.68	0.49	0.64	0.70	0.77	0.64	-	0.67	1.28
T_{all}	0.42	0.33	0.55	0.43	0.40	0.45	0.49	0.50	0.43	-	0.44	0.79
p	**0.00**	0.17	**0.01**	**0.03**	0.12	0.17	**0.01**	**0.00**	**0.00**	-	**0.01**	**0.01**
n	69	40	29	25	26	18	17	59	67	-	20	60
Baseline vs. T_2:												
B	0.61	0.39	0.91	0.68	0.54	0.62	0.69	0.76	0.62	-	0.69	1.30
T_2	0.44	0.33	0.59	0.43	0.46	0.42	0.60	0.52	0.46	-	0.60	0.81
p	**0.01**	0.22	**0.02**	0.09	0.31	0.10	0.38	**0.01**	**0.01**	-	0.37	**0.02**
n	64	37	27	24	23	17	16	54	60	-	16	53

Average Lane Incursion Time. The average duration of lane incursions decreased after the system was enabled, but this decrease was only significant overall for lane changes to the right as indicated in Table 17. Although there was not a reduction in the frequency of LCR conflicts shown in Table 16, there was a significant decrease in duration of the conflicts that occurred. There were also decreases in lane-incursion time for lane-change conflicts to the right, among women and younger drivers but this was only significant from baseline to T_{all}. For speeds above 55 mph, the decrease was significant from baseline to both treatment groups.

For lane changes to the left, the only significant decrease was amongst older drivers.

Table 17. Average lane incursion time for lane-change conflicts (seconds)

		Overall	Gender		Age (years)			Speed (mph)	
			Male	Female	20-30	40-50	60-70	25-55	55+
LCR	*Baseline vs. T_{all}:*								
	B	1.57	1.41	1.84	1.62	1.78	1.23	-	1.54
	T_{all}	1.06	1.08	1.03	0.81	1.07	1.48	-	1.11
	P	**0.02**	0.20	**0.03**	**0.02**	0.07	0.48	-	**0.03**
	N	35	22	13	15	11	9	-	34
	Baseline vs. T_2:								
	B	1.63	1.48	1.80	1.60	1.96	-	-	1.55
	T_2	0.98	0.88	1.11	0.79	1.32	-	-	1.00
	P	**0.02**	0.10	0.10	0.07	0.25	-	-	**0.02**
	N	27	15	12	12	9	-	-	26
LCL	*Baseline vs. T_{all}:*								
	B	1.08	1.13	1.00	0.92	1.35	0.81	1.59	1.24
	T_{all}	0.92	0.91	0.93	0.93	1.14	0.45	1.38	0.94
	P	0.28	0.26	0.77	0.98	0.47	0.28	0.65	0.14
	N	44	27	17	17	18	9	12	35
	Baseline vs. T_2:								
	B	1.00	1.11	0.84	0.92	1.21	0.77	1.61	1.22
	T_2	0.99	0.88	1.16	1.00	1.32	0.32	1.36	0.95
	P	0.94	0.17	0.33	0.77	0.73	**0.02**	0.73	0.19
	N	41	25	16	17	16	8	9	30

Maximum Lane Incursion Distance. When a car was more than 0.8 m into the next lane, the system would sometimes switch to tracking the lane dividers of the adjacent lane as the current lane. This switch meant that measures of lane-incursion distance over 0.8 m were unreliable and consequently were dropped from the analysis of lane-incursion distance (measures of lane-change conflict frequency, lateral acceleration and lane bust time were not affected by this error). This omission resulted in a decrease in the number of samples available for comparison and therefore a drop in statistical power: there was insufficient data to analyze lane changes to the right, and for lane changes to the left none of the comparisons that could be made were statistically significant (Table 42 in Appendix G).

2.3.4 Road-Departure Driving Conflicts

As with lane-change conflicts, events occurring while the vehicle was moving less than 25 mph were omitted for the following analyses because the lane tracking system is not operable at low speeds.

Conflict Rate. Road-departure conflicts were analyzed separately for straight and curved roads. For straight roads, there was no overall change in road-departure conflict rates, either for departures to the right or to the left (Table 43 in Appendix G). In terms of specific subgroups, the only significant result was for speeds over 55 mph: the rate decreased from 0.73 to 0.46 conflicts per 100 miles driven in T_{all} ($p = 0.03$, $n = 33$). To T_2, however, there was only a trend to significance: the conflict rate decreased from 0.79 to 0.56 ($p = 0.09$, $n = 25$).

Although the system did not have an effect on the frequency of road-departure conflicts while driving on straight roads, it did have a significant effect on drivers driving around curves. With the system enabled, drivers experienced overall decreases in road-departure conflicts on curves to both the right and the left as shown in Table 18. Furthermore, conflict rates dropped in all subgroups, although for departures to the right these drops were primarily significant for baseline to T_{all}. For departures to the left, there were significant decreases in all groups for T_{all}, and for T_2 rates dropped significantly for women, younger and middle-aged drivers, during night driving and in clear weather. For males and speeds over 55 mph, there was a trend towards significance.

Table 18. Average number of conflicts departing curved roads per 100 miles driven

Departing curved road to the right (CDR):

	Overall	Gender		Age (years)			Light		Weather		Speed (mph)	
		Male	Female	20-30	40-50	60-70	Night	Day	Clear	Adverse	25-55	55+
Baseline vs. T_{all}:												
B	0.54	0.49	0.61	0.51	0.50	0.65	1.58	0.61	0.57	-	1.03	0.78
T_{all}	0.41	0.40	0.41	0.41	0.38	0.44	0.57	0.47	0.42	-	0.70	0.53
p	**0.00**	0.09	**0.02**	0.15	0.13	**0.05**	**0.01**	**0.01**	**0.00**	-	**0.01**	**0.01**
n	71	41	30	24	29	18	27	59	69	-	44	45
Baseline vs. T_2:												
B	0.57	0.53	0.64	0.62	0.51	0.62	1.89	0.64	0.62	-	1.12	0.81
T_2	0.53	0.45	0.66	0.60	0.50	0.52	0.74	0.70	0.57	-	0.99	0.69
p	0.49	0.29	0.83	0.85	0.90	0.27	0.09	0.37	0.44	-	0.46	0.33
n	57	34	23	17	24	16	15	44	53	-	36	35

Departing curved road to the left (CDL):

	Overall	Gender		Age (years)			Light		Weather		Speed (mph)	
		Male	Female	20-30	40-50	60-70	Night	Day	Clear	Adverse	25-55	55+
Baseline vs. T_{all}:												
B	0.91	0.93	0.89	0.97	0.96	0.78	1.30	0.95	0.95	-	1.36	1.15
T_{all}	0.68	0.69	0.65	0.73	0.75	0.52	0.76	0.79	0.70	-	1.09	0.74
p	**0.00**	**0.00**	**0.00**	**0.02**	**0.01**	**0.01**	**0.00**	**0.01**	**0.00**	-	**0.00**	**0.00**
n	86	48	38	31	30	25	44	78	85	-	72	59
Baseline vs. T_2:												
B	0.93	0.93	0.94	1.01	0.98	0.78	1.32	0.97	0.96	-	1.35	1.14
T_2	0.72	0.77	0.65	0.77	0.78	0.60	0.94	0.86	0.74	-	1.22	0.89
p	**0.00**	0.07	**0.00**	**0.04**	**0.02**	0.14	**0.03**	0.16	**0.00**	-	0.21	0.09
n	83	48	35	29	29	25	40	73	82	-	67	47

Maximum Lateral Acceleration. As with lane-change conflicts, lateral acceleration could only be calculated for events occurring on straight roads. The results show no statistically significant overall effect (Table 44 in Appendix G). The only significance was a decrease in acceleration at speeds over 55 mph: maximum lateral acceleration fell from an average of 1.10 m/s^2 to 0.98 m/s^2 at T_2 ($p = 0.03$, $n = 25$).

Average Lane-Bust Time. Although there was no reduction in road-departure conflict frequency on straight roads, for departures to the right from straight roads there was a significant decrease in the duration of conflicts in terms of lane-bust time. This decrease was only significant from baseline to T_{all} (Table 19). In terms of effects within specific groups, there were significant decreases for T_2 for speeds between 25 and 55 mph for departures to the right, and for both males and younger drivers (aged 20-30 years) for departures to the left.

Table 19. Average lane bust time for straight road departures (seconds)

		Overall	Gender		Age (years)			Speed (mph)	
			Male	Female	20-30	40-50	60-70	25-55	55+
SDR	*Baseline vs. T_{all}:*								
	B	2.43	2.40	2.49	2.60	2.20	2.48	2.60	2.20
	T_{all}	2.15	2.15	2.14	2.23	2.10	2.11	2.26	2.08
	p	**0.05**	0.17	0.19	0.18	0.61	0.22	0.16	0.53
	n	52	35	17	19	17	16	29	33
	Baseline vs. T_2:								
	B	2.43	2.42	2.45	2.65	2.18	2.53	2.82	2.17
	T_2	2.26	2.24	2.31	2.45	2.24	2.11	2.12	2.32
	p	0.35	0.44	0.58	0.64	0.79	0.20	**0.01**	0.54
	n	42	31	11	12	16	14	21	25
SDL	*Baseline vs. T_{all}:*								
	B	2.21	2.21	2.21	2.12	2.40	2.11	2.09	2.40
	T_{all}	2.20	2.20	2.21	2.13	2.38	2.09	2.21	2.17
	p	0.94	0.92	1.00	0.96	0.92	0.92	0.45	0.12
	n	86	50	36	31	29	26	66	62
	Baseline vs. T_2:								
	B	2.24	2.25	2.23	2.16	2.46	2.11	2.16	2.41
	T_2	2.10	1.99	2.28	1.86	2.33	2.14	2.11	2.13
	p	0.19	**0.03**	0.75	**0.02**	0.54	0.87	0.79	0.07
	n	80	48	32	28	26	26	54	57

For curved roads, overall lane-bust durations decreased on average from baseline to T_2, but not significantly. The only significant changes were for women departing to the right, which increased from 1.84 to 2.19 seconds at T_{all} ($p = 0.05$, $n = 30$), and for men departing to the left, which decreased from 2.20 to 1.92 seconds at T_2 ($p = 0.04$, $n = 48$).

Maximum Lane-Bust Distance. As was the case for lane-change conflicts, for the analysis of road-departure lane-bust distance, departures in excess of 0.8 m were omitted (see above for explanation). The only significant overall change in lane-bust distance was an increase from 0.27 m at baseline to 0.32 m at T_{all} for departures to the left from straight roads ($p = 0.03$, $n = 73$) (Table 44 in Appendix G). Otherwise, there were some changes within various groups: lane-bust distance when departing straight roads to the right decreased in younger drivers (0.52 m to 0.25 m at T_{all}, $p = 0.01$, $n = 8$); older drivers increased lane-bust size in departures from straight roads to the left (0.23 m to 0.34 m at T_{all}, $p = 0.04$, $n = 23$); and women increased lane-bust size from curved roads to the right (0.18 m to 0.31 m at T_{all}, $p = 0.00$, $n = 21$, and 0.20 m to 0.30 m at T_2, $p = 0.02$, $n = 17$).

2.3.5 Curve-Speed Driving Conflicts

In the case of curve-speed driving conflicts, the system was disabled for vehicles entering a curve at less than 30 mph, and consequently those cases were excluded from the analyses. However, since we defined conflicts using data from the tightest point in the curve, and since drivers generally decelerate upon entering a curve, the speed during the conflict may be less than 30 mph. Consequently, in addition to the speed bins used for previous conflict types ("25–55 mph" and "55+ mph"), for curve-speed driving conflicts a third bin ("< 25 mph") was analyzed as well.

Conflict Rate. The number of conflicts where vehicles entered a curve driving too fast did not change significantly from baseline to either treatment period (Table 45 in Appendix G).

Maximum Lateral Acceleration. Lateral acceleration indicates the intensity of the conflict. There was an overall increase in lateral acceleration as compared to the baseline as seen in This increase was also significant within many categories: for T_{all} there was an increase for males, middle-aged drivers and speeds 25–55 mph; for T_2 there was an increase just for speeds 25–55 mph. However, for both treatment periods there was a significant decrease in lateral acceleration for speeds less than 25 mph.

The increase in lateral acceleration during curve-speed conflicts in the middle speed bin could be due to drivers' overall increase in lane positioning, as discussed previously in Section 2.2.4. These results are also supported by the decrease in lane departure warnings discussed in Section 4.2.3. If a driver makes an effort to stay within their lane in a curve, they would experience a higher lateral acceleration than if they were to take a wider trajectory around the curve by departing their lane.

Speed at CPOI. The amount of excess speed with which drivers entered curves was defined as the speed at the curvature point of interest (CPOI, the calculated midpoint of the curve) minus the maximum safe speed (calculated via a method outlined in Lam, et al., 2009). Although overall there was a slight increase in this average speed excess, none of the changes were statistically significant (Table 46 in Appendix G).

Table 20. Average maximum lateral acceleration (m/s^2) in curve-speed conflicts

	Overall	Gender		Age (years)			Speed (mph)		
		Male	Female	20-30	40-50	60-70	<25	25-55	55+
Baseline vs. T$_{all}$:									
B	4.16	4.24	4.03	4.24	4.10	4.14	4.87	4.12	4.23
T$_{all}$	4.25	4.35	4.05	4.32	4.20	4.20	4.66	4.22	4.36
p	**0.02**	**0.02**	0.52	0.17	**0.04**	0.42	**0.03**	**0.01**	0.62
n	75	48	27	29	28	18	18	73	18
Baseline vs. T$_2$:									
B	4.17	4.24	4.04	4.24	4.12	4.14	4.91	4.13	4.26
T$_2$	4.26	4.35	4.09	4.34	4.18	4.24	4.59	4.24	4.47
p	**0.04**	0.05	0.42	0.18	0.22	0.31	**0.00**	**0.02**	0.52
n	73	48	25	29	26	18	14	71	14

2.3.6 Driver Attention in Driving Conflicts

As in the analysis of attention in overall driving discussed in Section 2.2.6.1, these results were obtained through video analysis. This analysis examines driver engagement in secondary tasks and the likelihood that the driver's eyes are off the forward scene in conflict scenarios. These results address whether or not drivers are more likely to have diverted their attention when conflict scenarios occur than in overall alert scenarios. This analysis includes data from all valid analyzed alerts that were triggered by a conflict scenario.

Figure 12 shows the proportion of conflicts where drivers engaged in secondary tasks. Across conflict types, drivers were engaged in secondary tasks in 63 percent of alerts that led to conflicts, compared to 55 percent of all alerts analyzed, suggesting that drivers are more likely to get into a conflict scenario when engaging in secondary tasks. Secondary tasks were more common during road-departure conflicts than during other conflict types. Many times when drivers were engaged in secondary tasks, they accidently drifted out of their lane, which was likely to cause a road-departure conflict.

The results for the second metric of driver attention, eyes-off-forward-scene, are illustrated below in Figure 13. Similar to the results for secondary tasks, drivers were more likely to have their eyes off the forward scene during an alert that led to a conflict scenario (10 percent of alerts issued for conflict scenarios) than in all analyzed alerts (seven percent). Drivers were least likely to have their eyes off the forward scene for curve-speed conflicts, since they generally kept their eyes on the road while negotiating curves.

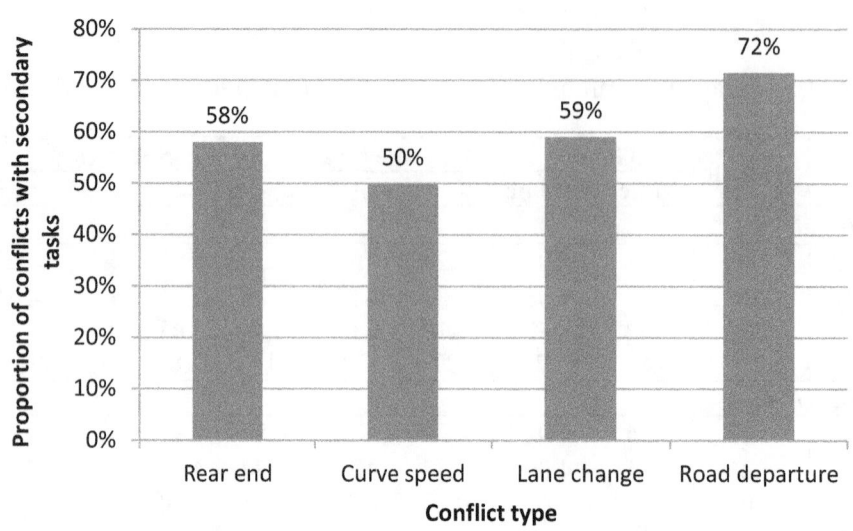

Figure 12. Proportion of valid alerted conflicts with secondary tasks by conflict type

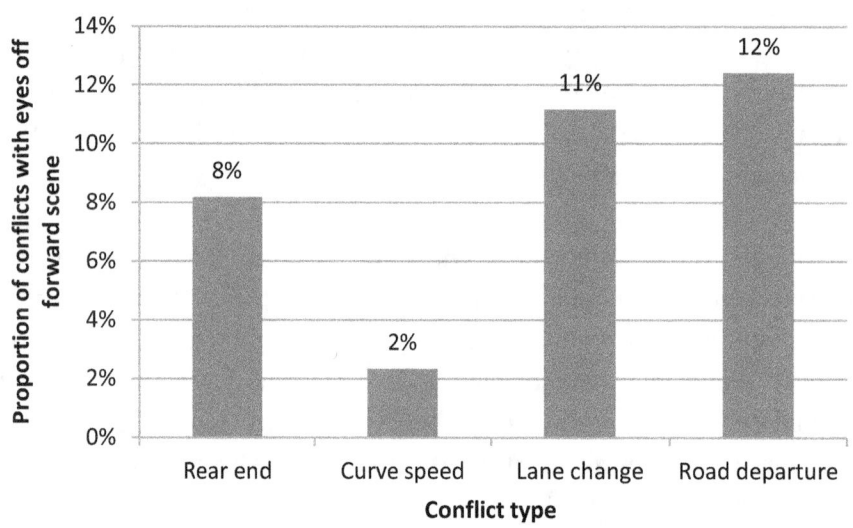

Figure 13. Proportion of valid alerted conflicts with drivers' eyes off forward scene

2.4 Near Crash Experiences

The analysis of near crashes addresses driving conflicts that resulted in a driver response above a certain intensity level. Thus, near crashes constitute a subset of longitudinal and lateral driving conflicts in which an intense driver response was observed during the field test data based on various kinematic measures. Near-crash thresholds were determined using distributions of intensity measures recorded in the field test. As in the previous sections, the frequency of near crashes and driver attention behavior leading up to near crashes were examined between treatment periods conditions by age, gender and driving speed range. The near-crash thresholds for each conflict type and the number of near crashes for each can be found in Appendix H.

By applying the near crash criteria, the query of the processed numerical database yielded 1,946 potential near crashes. A video analysis was conducted for each near crash to determine whether or not a valid threat was actually present in the driving scenario. Of these cases, a total of 1,810 or about 93 percent contained a valid threat. Rear-end (RE), curve-speed (CS), and road-departure (RD) near crashes were accurately detected by the data mining algorithms as shown in Table 21. The identification of lane-change/merge near crashes was less accurate than the other near crash types due to difficulties in determining whether the adjacent target was a vehicle or a stationary object such as a guard rail.

Table 21. Breakdown of near crashes and their validity rate

Near Crash Type	Number of Near Crashes	% Valid Near Crashes
Rear End	370	99.7%
Curve Speed	461	97.1%
Lane Change/Merge	232	68.4%
Road Departure	883	94.8%
Total	1,946	93.1%

The potential effect of the LDW function on opposite-direction crashes caused by an unintentional drift by the host vehicle into an adjacent lane of oncoming traffic was also examined. This analysis was conducted using a sample of videos showing driving episodes in which LDW-C or LDW-I alerts were triggered, as discussed in Section 1.4.2. Driver performance was compared between baseline and treatment conditions for all drivers as well as separately for each age and gender group. The following measures were used to characterize driver performance:
- Proportion of alerts on road edges without adjacent lanes of opposite-direction traffic
- Proportion of alerts for adjacent lanes with oncoming traffic in which a vehicle approached the host vehicle from the opposite direction
- Time-to-collision, measured by reviewing the videos from the time of the alert onset until the overlap of the fronts of the two vehicles

2.4.1 Exposure to Near Crashes

Driver involvement in near crashes was analyzed using the exposure measure of the number of near-crash encounters per 1,000 miles traveled. Paired *t*-tests did not show any statistically significant differences in the exposure to rear-end and curve-speed near crashes.

Table 22 shows the statistically significant effect of the integrated system on the rate of valid near crashes overall and by age and gender based on the results of paired t-tests. For all near crash types combined, only younger drivers showed a significant change in near-crash exposure with the system enabled (19 percent reduction). Younger drivers also showed a significant reduction in LCM and RD near crashes individually. Overall, drivers showed a 33 percent reduction in LCM near crashes and 19 percent reduction in RD near crashes. Paired t-tests did not show any statistically significant differences in the exposure to rear-end and curve-speed near crashes.

Table 22. Paired t-test results of average number of near crashes per 1,000 miles driven

	Overall	Gender		Age (years)		
		Male	Female	20-30	40-50	60-70
All Near Crashes						
B	9.64	10.64	8.31	12.20	9.24	7.06
T_{all}	9.19	10.00	8.10	9.84	9.82	7.74
p	0.45	0.42	0.82	**0.05**	0.57	0.44
n	91	52	39	33	30	28
LCM						
B	2.12	1.72	2.73	2.63	1.92	1.55
T_{all}	1.43	1.08	1.93	1.48	0.95	1.79
p	**0.02**	**0.03**	0.20	**0.02**	0.06	0.58
n	37	22	15	16	10	11
RD						
B	5.40	5.45	5.34	6.19	5.12	4.77
T_{all}	4.38	4.62	4.05	3.99	5.02	4.19
p	**0.00**	0.08	**0.03**	**0.00**	0.87	0.33
n	74	43	31	27	24	23

Paired t-test results for near-crash rates overall and for each near crash type at speeds between 25 and 55 mph and at speeds over 55 mph are shown in Table 23. While drivers experienced higher rates of near crashes in the lower speed bin, the system was more effective at reducing near crashes in the higher speed bin. Drivers experienced a 32 percent reduction in all near crashes that occurred above 55 mph, with significant reductions for both types of near crashes analyzed. Rear-end and curve-speed near crashes were very rare at speeds over 55 mph. In the lower speed bin, drivers experienced a 21 percent reduction in road-departure near crashes.

Road-departure near crashes were also analyzed by departure direction and by age and gender groups. Rates of road-departure near crashes were much higher to the left than to the right as indicated in Table 24. The reduction in RD near crashes to the right was significant for all drivers combined. Younger drivers, although they showed a reduction in road-departure near crashes to the right, were the only group of drivers who did not show a statistically significant reduction. However, younger drivers did show a significant decrease in road-departure near crashes to the left, as did older drivers and males.

Table 23. Paired *t*-test results of near-crash rates by speed bin and near crash type

	Overall	RE	CS	LCM	RD
25-55 mph					
B	15.27	6.01	11.46	-	9.46
T_{all}	14.39	4.87	10.30	-	7.43
p	0.34	0.18	0.27	-	**0.01**
n	73	24	29	-	50
55+ mph					
B	9.84	-	-	4.20	7.17
T_{all}	6.66	-	-	2.52	4.98
p	**0.00**	-	-	**0.00**	**0.01**
n	66	-	-	22	47

Table 24. Paired *t*-test results of road-departure near-crash rates by direction and age/gender groups

	Overall	Gender		Age (years)		
		Male	Female	20-30	40-50	60-70
Left						
B	4.58	4.46	4.79	5.76	4.27	3.60
T_{all}	3.69	3.39	4.23	3.59	4.72	2.47
p	0.06	**0.01**	0.62	**0.004**	0.65	**0.04**
n	62	40	22	21	23	18
Right						
B	2.75	2.43	3.24	2.76	2.43	3.02
T_{all}	1.68	1.39	2.10	1.47	1.41	2.07
p	**0.00**	**0.01**	**0.04**	0.10	**0.03**	0.05
n	35	21	14	11	11	13

Nonparametric statistical sign tests were also performed to supplement the results of paired *t*-tests. The sign test uses only the sign or direction of differences between pairs of observations in the paired-sample case, and does not take into consideration the magnitude of these differences. The treatment condition led to lower near-crash rates than the baseline condition with 93 percent confidence level (two-tail $p = 0.07$) in rear-end near crashes, 76 percent confidence level in curve-speed near crashes, 96 percent confidence level in left road-departure near crashes, and 99 percent confidence level in right road-departure near crashes.

Based on the analysis of 9,089 videos associated with LDW-C and LDW-I alerts issued for left lateral drifts, the vehicle was on an undivided roadway drifting onto an adjacent, opposite direction traffic lane in only 29 percent of the cases. In 12 percent of these opposite-direction lane departure cases, another vehicle was approaching in the opposite-direction lane. The time that it would take for the two vehicles to meet from the onset of the LDW alert was also determined for those cases when the opposite-direction lane was occupied. This time was about three seconds or more (estimated minimum response time required to avoid a collision including system warning delay, average driver response time, and vehicle response) in 45 percent of these cases. Thus, an assumption could be made that a left lateral drift warning may have the potential to prevent an opposite-direction crash in 45 percent of the cases when a vehicle drifts to an occupied lane with opposite-direction traffic.

2.4.2 Driver Attention in Near Crashes

Figure 14 illustrates driver engagement in secondary tasks during alerted near crash events. For all near crash types, drivers engaged in secondary tasks in 68 percent of alerted near crash scenarios compared to 63 percent of alerted conflicts and 45 percent of alert scenarios, indicating that drivers are more likely to be engaged in secondary tasks in alert scenarios that lead to near crashes. Whereas in the conflict analysis road departure had the highest rate of secondary tasks (72 percent), rear-end near crashes have the highest rate of secondary tasks. Similarly, drivers had higher rates of eyes-off-forward-scene behavior during near crashes than during conflicts for all alerts (13 percent compared to 10 percent and 7 percent, respectively). As shown in Figure 15, drivers also had the highest proportion of eyes-off-forward-scene behavior for rear-end near crashes, suggesting that when drivers are engaged in a secondary task or their eyes are off the road during a rear-end conflict it is more likely to turn into a severe event than for other conflict types.

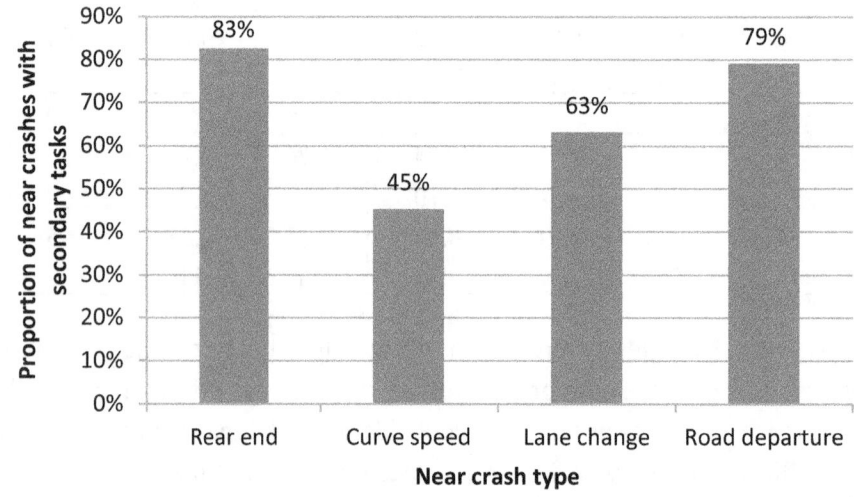

Figure 14. Proportion of near crashes where drivers were engaged in secondary tasks

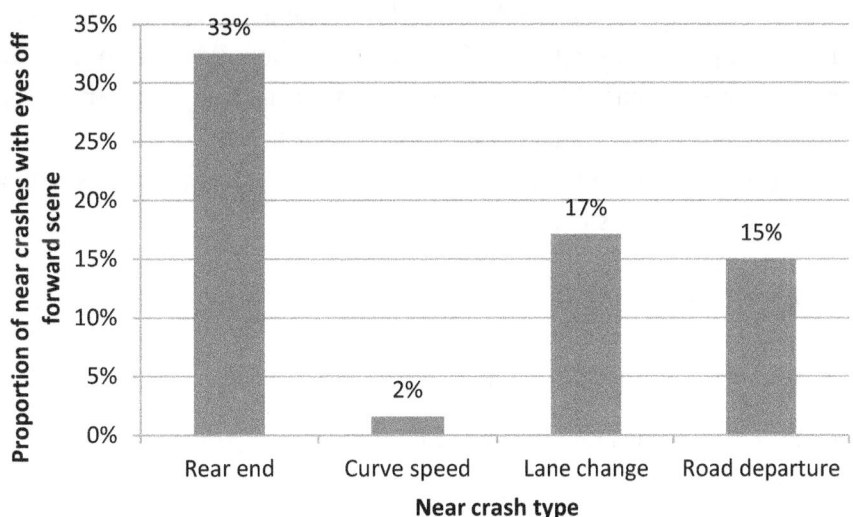

Figure 15. Proportion of near crashes where drivers had their eyes off the forward scene

2.5 Projection of Potential Safety Benefits

This analysis projects the potential safety benefits of the integrated safety system in terms of the annual frequency of target crashes that might be avoided with full deployment of the system, N_a, where:

$$N_a = \sum_{i=1}^{n} N_{wo}(S_i) \times E(S_i) \qquad (1)$$

$n \equiv$ Number of applicable pre-crash scenarios, S_i

$N_{wo}(S_i) \equiv$ Annual number of target crashes preceded by S_i prior to system deployment

$E(S_i) \equiv$ System effectiveness in avoiding target crashes preceded by S_i

Values of $N_{wo}(S_i)$ are obtained from the GES as listed in Table 1. $E(S_i)$ is expressed as:

$$E(S_i) = 1 - \frac{P_w(C|S_i)}{P_{wo}(C|S_i)} \times \frac{P_w(S_i)}{P_{wo}(S_i)} \qquad (2)$$

$P_w(C|S_i) \equiv$ Probability of a crash in treatment given an S_i encounter

$P_{wo}(C|S_i) \equiv$ Probability of a crash in baseline given an S_i encounter

$P_w(S_i) \equiv$ Probability of an S_i encounter in treatment

$P_{wo}(S_i) \equiv$ Probability of an S_i encounter in baseline

The ratios $\dfrac{P_w(C|S_i)}{P_{wo}(C|S_i)}$ and $\dfrac{P_w(S_i)}{P_{wo}(S_i)}$ are known respectively as the crash prevention ratio (PR) and scenario exposure ratio (ER).

The experience of near crashes in the baseline and treatment test conditions provides a good measure to estimate the potential safety benefits because it captures the frequency and severity of driving conflicts encountered during the field test. Thus, near-crash rates serve as surrogate measures for the crash prevention (PR) and scenario exposure (ER) ratios presented above. Equation (2) is rewritten below to incorporate driver exposure to near crashes with and without the assistance of the integrated system:

$$E(S_i) = 1 - PNC_w(S_i)/PNC_{wo}(S_i) \qquad (3)$$

$PNC_w(S_i) \equiv$ Near-crash rate of type S_i in treatment
$PNC_{wo}(S_i) \equiv$ Near-crash rate of type S_i in baseline

Equation (3) was applied to individual drivers who experienced at least one near crash in baseline and treatment conditions. Figure 16 illustrates the descriptive statistics of individual effectiveness values in various near crashes in terms of the average and 95 percent confidence interval. Values shown in each bar refer to the number of subjects who met the criterion of exposure to near crashes in both test conditions. The 95 percent confidence intervals are positive for system effectiveness in rear-end, lane-change/merge, all road-departure, left road-departure, and right road-departure near crashes.

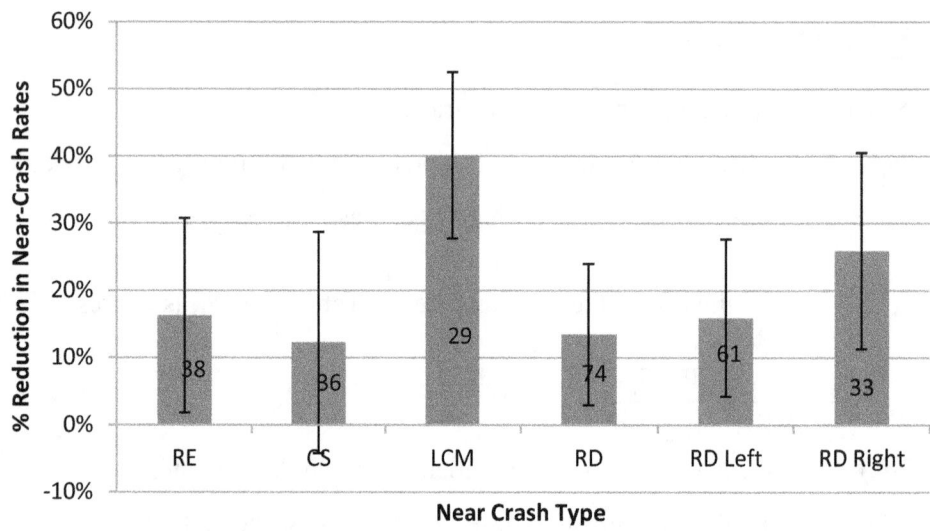

Figure 16. Average system effectiveness values in various near crashes

Potential safety benefits of the integrated safety system were projected using system effectiveness values, $E(S_i)$ shown in Figure 16, and annual crash frequencies, $N_{wo}(S_i)$ listed in Table 1. Table 25 shows the estimated effectiveness values based on 95 percent confidence bounds and resulting crash reductions for each system function and for the integrated system.

With an overall system effectiveness range between six and 29 percent, approximately 162,000 to 788,000 police-reported crashes could be prevented annually if all light vehicles in the United States were equipped with the integrated safety system. The following list ranks the system functions in terms of their maximum annual crash reduction potential:

1. FCW: 450,000 police-reported rear-end crashes
2. LCM: 163,000 police-reported lane-change crashes
3. LDW-C right: 101,000 police-reported road-departure crashes
4. LDW-C left: 47,000 police-reported road-departure and opposite-direction crashes
5. LDW-I: 27,000 police-reported lane-change crashes

Safety benefits cannot be estimated for the CSW function and for the LCM function in turning scenarios due to the lack of statistically significant differences in the mean values of exposure to near crashes recorded during the field test.

Table 25. Crash reduction estimates with full deployment of the integrated system in light vehicles

Function	Pre-Crash Scenario	Annual Target Crashes	Estimated Crash Reduction		Estimated System Effectiveness	
			Minimum	Maximum	Minimum	Maximum
FCW	Rear-end/lead vehicle stopped	1,462,000	27,000	450,000	2%	31%
	Rear-end/lead vehicle decelerating					
	Rear-end/lead vehicle moving					
CSW	Negotiating a curve/lost control	181,000	Insufficient data to estimate			
LCM	Changing lanes/same direction	311,000	86,000	163,000	28%	53%
	Turning/same direction	195,000	Insufficient data to estimate			
LDW-I	Drifting/same direction	51,000	14,000	27,000	28%	53%
LDW-C Right	Right road departure/no maneuver	249,000	28,000	101,000	11%	41%
LDW-C Left	Left road departure/no maneuver	122,000	5,000	34,000	4%	28%
	Opposite direction/no maneuver	103,000	2,000	13,000	2%	12%
Integrated System	All	2,674,000	162,000	788,000	6%	29%

3. Driver Acceptance

The second goal of the independent evaluation deals with driver acceptance, which was assessed in terms of the following five objectives:

- Ease of use: determine the usability of the integrated safety system
- Perceived usefulness: analyze drivers' subjective assessments of safety using the integrated safety system
- Ease of learning: assess how well drivers understand the system
- Advocacy: determine the drivers' expressed willingness to drive a truck equipped with the integrated safety system
- Driving performance: monitor whether system use leads to unintended consequences, as well as any behavioral adaptations.

> **HIGHLIGHTS**
> - Eighty-two percent or drivers felt that the system would increase their driving safety.
> - Drivers' favorite feature of the integrated system was the blind spot monitors.
> - Only one third of drivers said that the integrated system issued nuisance warnings too frequently.
> - Only seven drivers reported negative behavior adaptations when driving with the integrated system.
> - Older drivers found the system to be more useful than the other age groups.
> - Drivers found the lateral warning systems to be more useful and more desirable than the longitudinal warnings.
> - Drivers reported exposure to false warnings was consistent with their actual exposure.

This section presents notable results from the driver-acceptance analysis based on survey data by age and gender group. It also includes the results of driver acceptance broken down by demographic and system performance variables.

3.1 Driver Acceptance Technical Approach

Driver acceptance was assessed by using subjective data in the form of survey responses. The data were quantified overall as well as separately by independent variables related to drivers' demographic information and experience with the integrated system. This section discusses the measures used to define acceptance, as well as the independent variables and methodology used in these analyses.

3.1.1 Acceptance by Driver and Objective

The five objectives of driver acceptance were rated subjectively by each test participant. Raw subjective data consist of numerical survey responses, written survey responses, verbatim comments, and results of the debriefing interview.

Most items on the post-drive survey asked drivers to rate various items on a seven-point scale with anchored points ranging from "strongly disagree" to "strongly agree." All survey questions can be found in Appendix A. A score of one to three indicate a negative response while five to seven indicate a positive response. An answer of four indicates a neutral response. Because the interpretation of the scale is somewhat dependent upon the participant, the independent evaluation reports all driver responses in terms of positive, neutral, or negative, rather than through numerical values. The meaning of a number six, for example, may vary from driver to driver; but overall, ratings of above four indicate positive feelings, values below four indicate negative feelings, and a response of four is considered neutral. Quantifying survey data in this manner removes some of the scaling subjectivity of the data.

Each survey item was mapped to a driver-acceptance objective as shown in Appendix I. For each driver, responses to surveys mapped to a given objective were combined for an overall percent positive, negative, or neutral response. Overall results for an objective across drivers were in the form of proportions of positive responses. Open-ended survey responses and verbatim comments were quantified in terms of frequency of responses across drivers.

Three focus groups were conducted during the light vehicle field test. All 108 drivers were invited to participate in the focus groups and total of 31 drivers attended one of the three focus groups. Focus groups were held in a round table discussion style where the moderator asked the group a total of 25 questions over the course of the two hour session. Results of the focus group are expressed in terms of frequency of response, or as anecdotal comments.

3.1.2 Acceptance by Independent Variables

Demographic and driving history data are used to determine if any driver characteristics affected driver acceptance of the integrated system. Driver acceptance data were also assessed according to drivers' actual experiences with the integrated safety system to provide insight into whether or not the type and frequency of alerts received by drivers influenced their perception of the system.

3.1.2.1 Demographic Variables

Demographic and driving history includes characteristics of the driver and their driving patterns. This information was obtained through a pre-drive survey that collected driver demographic information. Each driver completed this survey at the beginning of their participation in the field test.

3.1.2.2 Driver Experience Variables

The variables of the driver experience represent metrics about the types of alerts the drivers received while driving with the integrated safety system, the frequency of alerts, alert validity, driving patterns, and conflict rates received during their participation in the field test. These variables are important because the experience a driver has with the system and the type of driving that they do with the instrumented vehicle can have an impact on their acceptance of the warnings. All experience metrics refer only to the system performance in the treatment period since the performance during this time period was the basis of drivers' subjective responses. All alert and conflict rates refer to the number of alerts per 100 miles. The 26 variables used in the analysis of driver acceptance by driver experience are listed in Table 26.

Table 26. Driver experience categories used in driver-acceptance analysis

Category	Variable
Alert numbers	All alerts
	FCW
	CSW
	Side imminent
	LDW-C
Alert rates	All alerts
	FCW
	CSW
	Side imminent
	LDW-C
False alert rates	All alerts
	FCW
	CSW
	Side imminent
	LDW-C
Driving patterns	Treatment mileage
	Average trip length
	% of daytime/nighttime driving
	% freeway/non-freeway driving
Conflict rates	Rear end
	Curve speed
	Lane change
	Road departure
	All conflicts

3.2 Subjective Results

General results as well as results within each driver-acceptance objective are presented based on survey responses, feedback from focus groups, and verbatim comments from the debriefing interviews. Results are broken down by demographic variables where effects were observed.

3.2.1 General Feedback

Drivers' responses to the open-ended question, "What did you like most about the integrated system?" are illustrated in Figure 17. Over half of the drivers reported that their favorite element of the system was the blind spot monitor lights. Drivers generally found these to be convenient, accurate, and not obtrusive. Younger drivers more frequently reported favoring the drift warnings, and older drivers were more likely to report that an increase in safety was their favorite thing about the system. Women were more likely to report liking the blind spot monitor lights, and men were more likely to report liking the drift warnings.

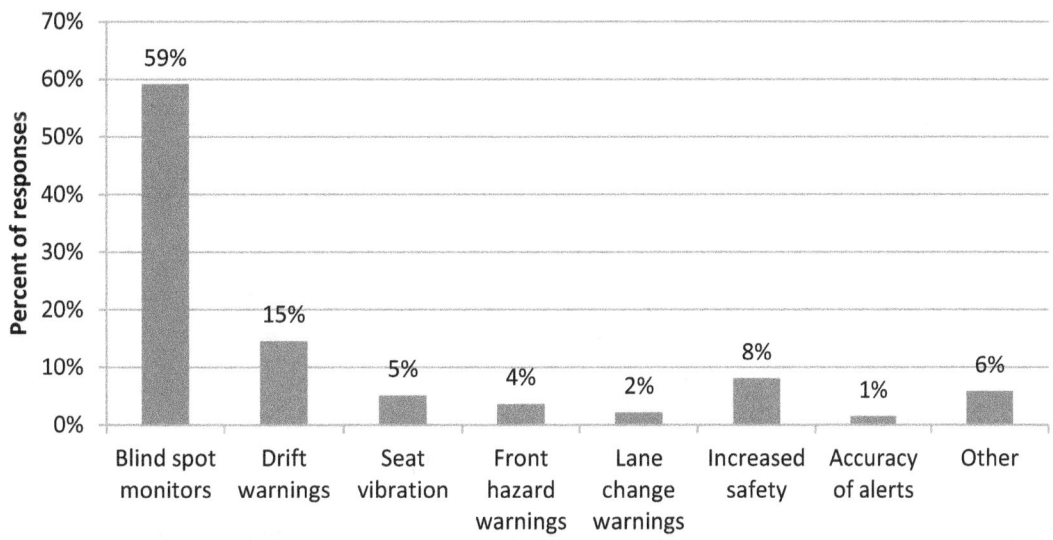

Figure 17. System features liked best by drivers

Figure 18 illustrates drivers' responses to the open-ended question, "What did you like least about the integrated system?" Half of the responses mentioned false warnings. Of these responses, 14 drivers specifically mentioned the false drift warnings, nine mentioned the false forward-collision warnings, and eight did not like the false side-hazard warnings. Many drivers specifically mentioned that they disliked when they received warnings when there was no threat present (compared to an unnecessary true positive). The least favorite element of the system for 10 percent of drivers was the curve-speed warning system. Drivers commented both that they received alerts when no curve was present, and that they disliked receiving warnings when they were familiar with a curve and how fast they could safely drive it.

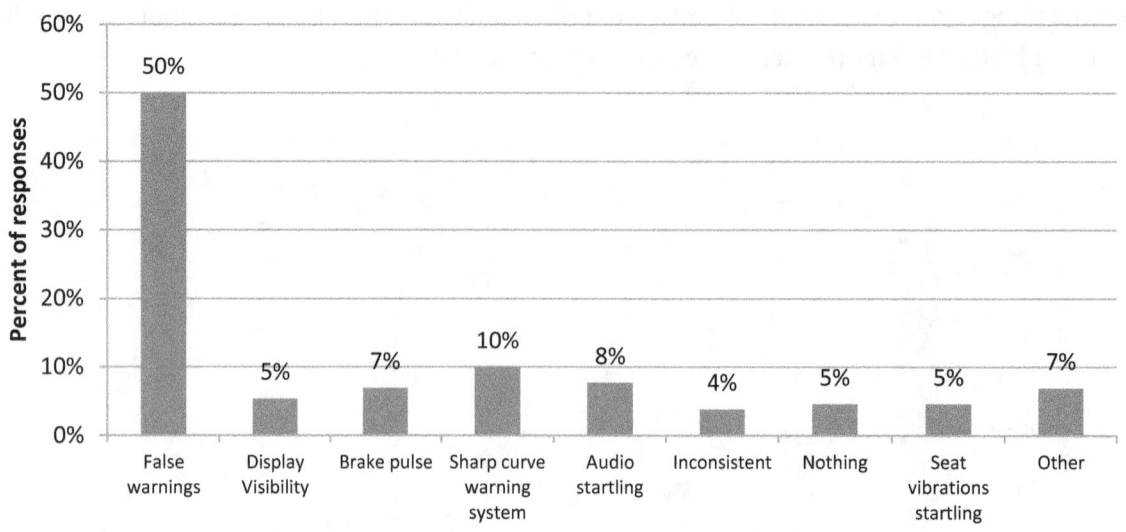

Figure 18. System characteristics liked least by drivers

When asked in which situations they found the integrated system to be most helpful, most drivers said that they found the system to be most useful when checking for vehicles in their blind spot. As illustrated in Figure 19, 24 percent of drivers thought the system was most helpful when they were drifting out of their lane. A few drivers mentioned that the system was helpful when they were distracted or fatigued. One driver commented, "the system was helpful because it made me realize how much I was distracted by phone calls (due to drift warnings)."

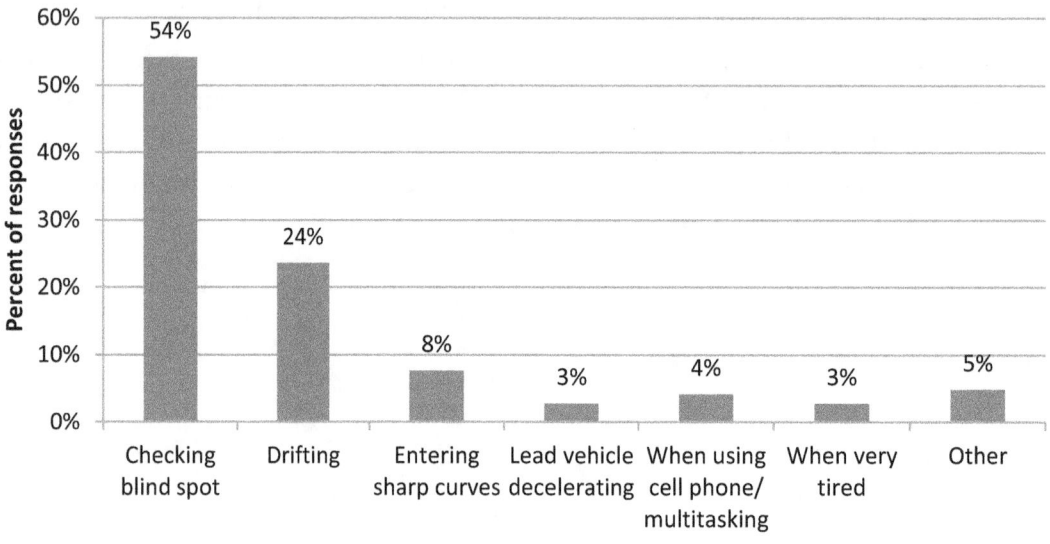

Figure 19. Situations in which drivers found the integrated system to be most helpful

3.2.2 Ease of Use

Overall results of the 18 numerical style questionnaire items addressing ease-of-use are shown in Figure 20 and Figure 21. These figures show the distribution of the percent of the 18 ease-of-use questionnaire items drivers responded to positively. All but 10 drivers responded positively to at least 60 percent of the ease-of-use questions, and 19 drivers responded positively to all 18 questionnaire items. There were no pronounced differences in drivers' opinions of ease-of-use of the integrated system between age groups or genders.

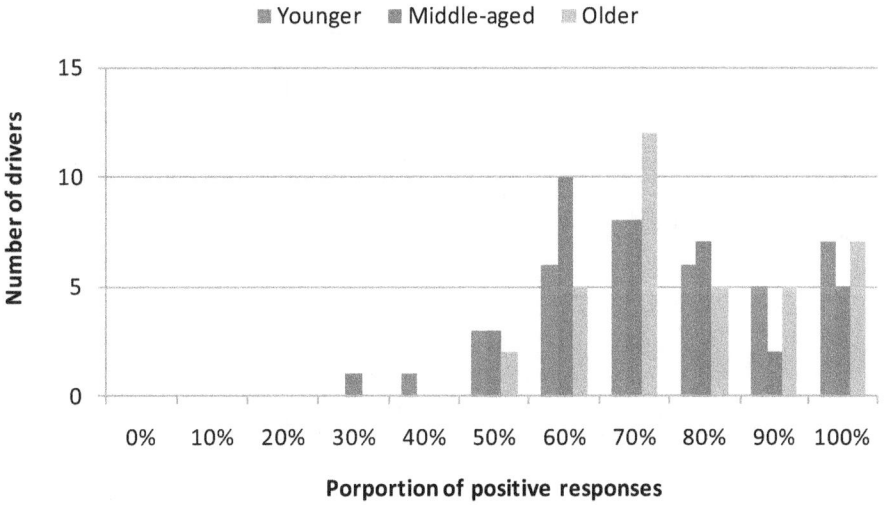

Figure 20. Distribution of ease-of-use responses by age group

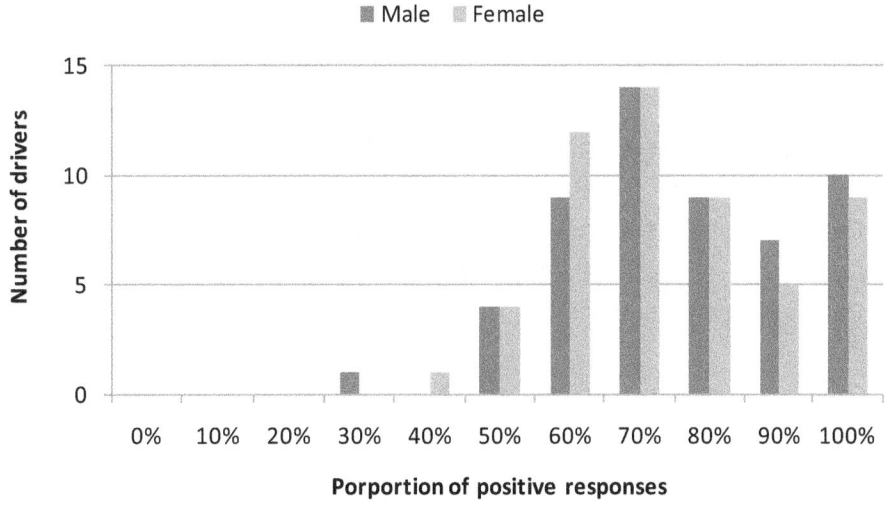

Figure 21. Distribution of ease-of-use responses by gender

Figure 22 illustrates how drivers responded to the statement, "the integrated system made driving easier." The intent of this questionnaire item is to understand how the integrated system affects drivers' workloads while driving, a positive response indicating that the driver felt that driving with the system reduced the workload of driving. The top portion of Figure 22 shows the results broken down by age group. With positive responses from 78 percent of the drivers, older drivers were the most likely to feel that the integrated system made driving easier. Less than half of younger drivers felt that the driving with the integrated system reduced their workload. The bottom portion of Figure 22 shows the breakdown of responses by gender. Female drivers were slightly more likely than males to think that the system made driving easier.

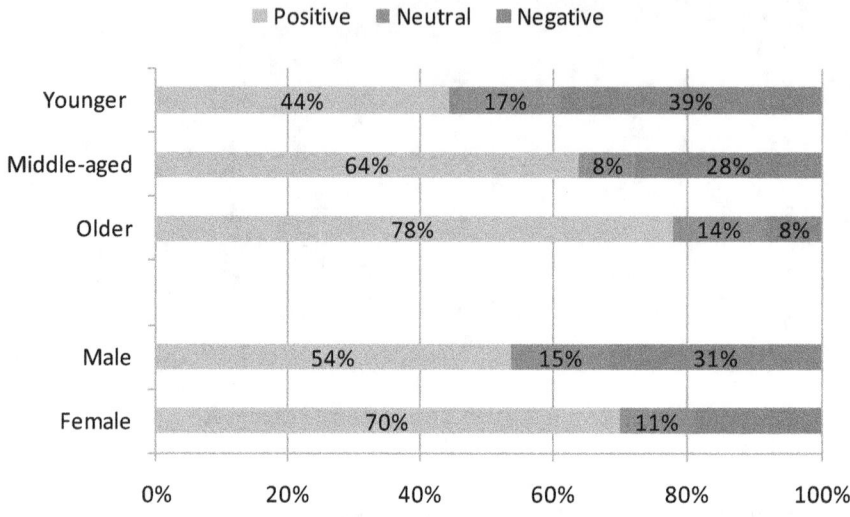

Figure 22. Responses to the statement, "the integrated system made driving easier"

Similar to the trend shown above, there was a slight effect of the number of years of driving experience on drivers' perception of whether or not the integrated system made driving easier. Contrary to some drivers' comments that the integrated system would be a "helpful learning tool for newer drivers," drivers with fewer than 10 years of experience rated the helpfulness of the integrated system the lowest. Drivers with 40-50 years of driving experience agreed most strongly that the integrated system made driving easier. The average questionnaire response by years of driving experience is shown in Figure 23.

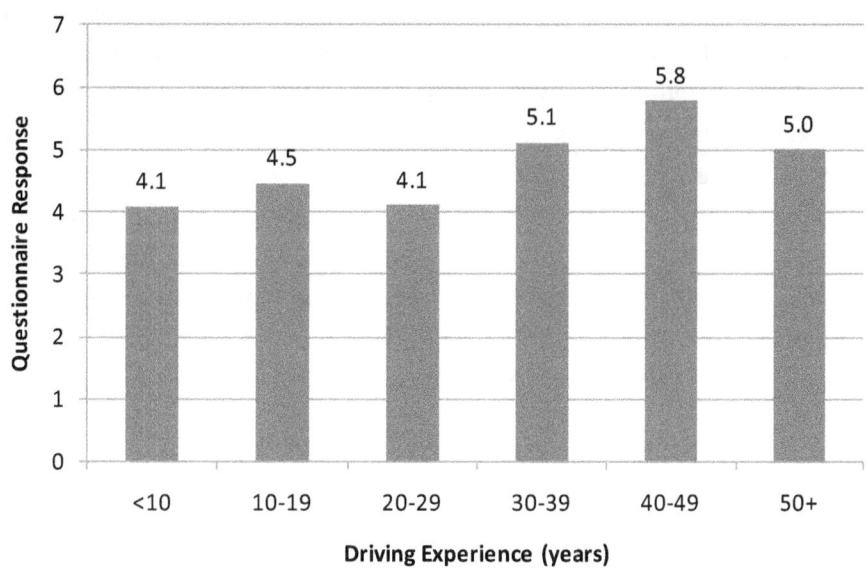

Figure 23. Responses to the statement, "the integrated system made driving easier" by years of driving experience

Drivers' responses about predictability and consistency of the integrated system are broken down by age group and gender in Figure 24. Predictability of the warnings was important for drivers' understanding of why warnings were issued as well as their trust in the system. Overall, 67 of the drivers felt that the system was predictable and consistent.

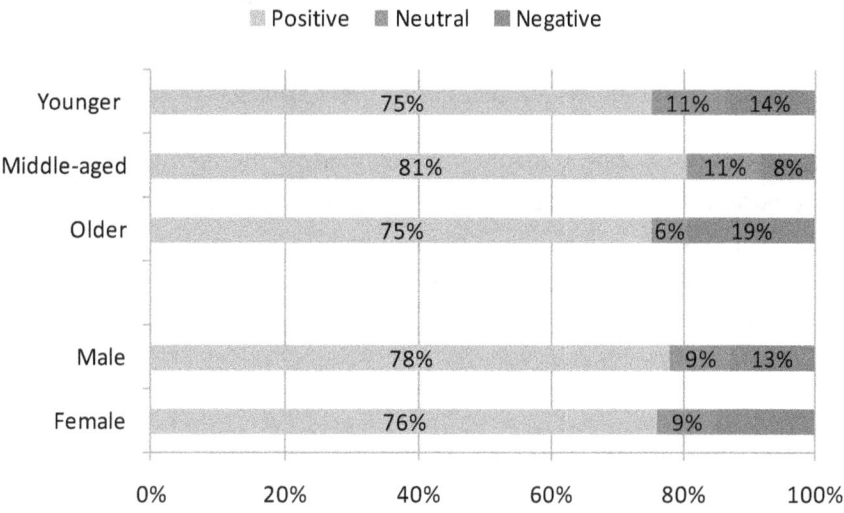

Figure 24. Responses to the statement, "the integrated system was predictable and consistent"

For each type of warning, drivers were asked the question, "I always understood why the integrated system provided me with a warning." A negative response to one of these questions indicated that the driver did not have a good mental model of how the alert type was supposed to work. Lack of understanding of the warnings could be due to inconsistencies in the timing of the warning or false alarms. Figure 25 shows drivers' reported understanding overall and by each warning type. Sixty percent of drivers said that they always understood why the system provided them with a warning while 29 percent of drivers said that they did not understand. The highest rate of negative responses was for the brake pulse warning with 32 percent, followed by the auditory warnings with 23 percent. Auditory warnings were issued for forward-collision, curve-speed, and side-hazard alerts. Seat vibration warnings (cautionary lane departure alerts) and blind spot indicators (an element of the side-hazard warnings) showed very high understanding by drivers indicating high accuracy and consistency for these warning types.

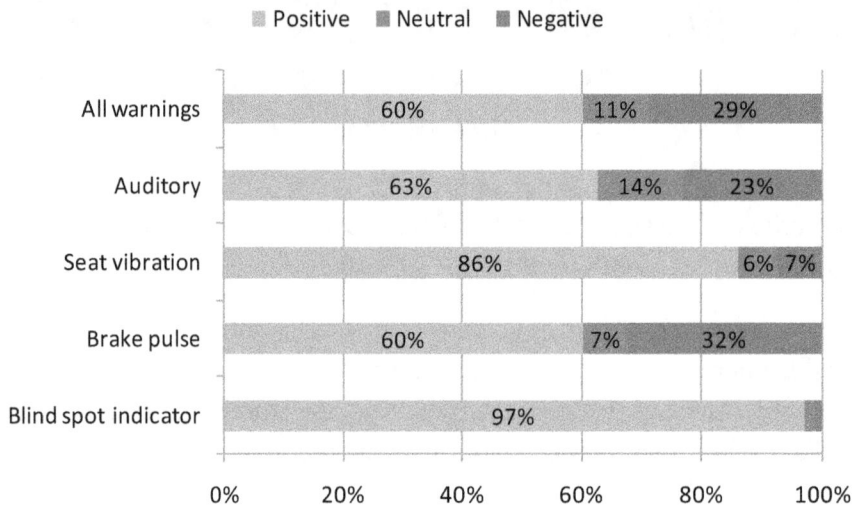

Figure 25. Drivers' understanding of the warnings by warning type

There was a slight effect of education level on understanding the warnings. Drivers with higher education levels reported lower understanding of the warnings overall, of auditory warnings, and of the brake pulse than drivers with a high school education. Figure 26 shows the average questionnaire response for drivers within each education level. A lower response indicates lower understanding of the warnings. This effect of education levels may be due to more educated drivers being more critical of the system.

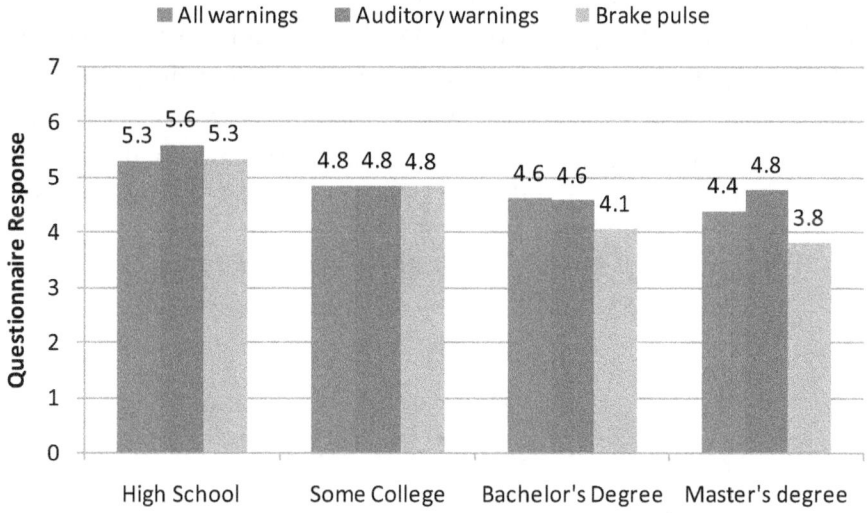

Figure 26. Understanding of warnings by warning type and education level

Figure 27 illustrates responses to the statement, "the warnings were not annoying" by alert type. A positive response to this statement indicates agreement that the warnings are not annoying and a negative response indicates that the driver was annoyed by the warnings. Overall, younger and middle-aged drivers were more likely to report annoyance with warnings than older drivers. The brake pulse warnings showed the highest rate of annoyance compared to the other alert types. In the focus group, most drivers mentioned that they would like to remove the brake pulse from the forward warnings. Eighteen of the drivers reported that they found the auditory warnings annoying, and when asked what they would change about the system, 11 of the drivers suggested that the sound used for audio warnings be changed to something less startling.

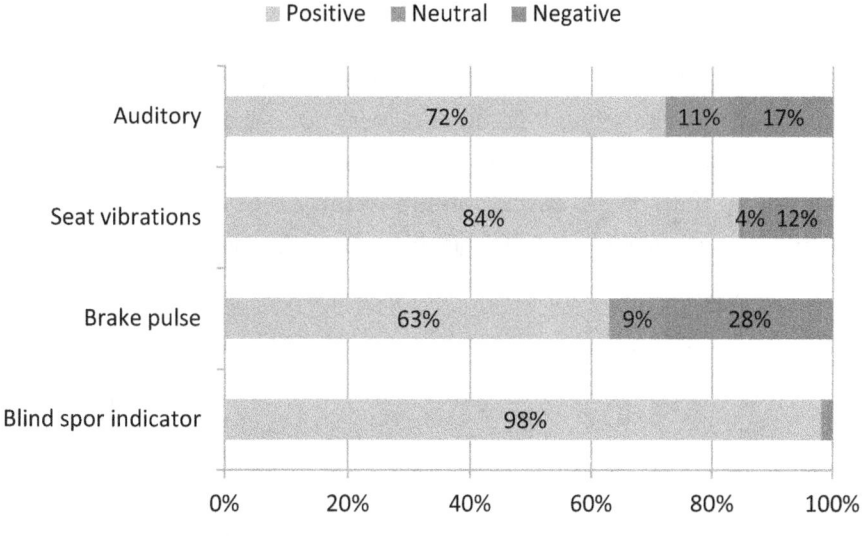

Figure 27. Drivers' responses to the survey item, "The alerts were not annoying"

As with understanding of warnings, there was a slight effect of education level on reported annoyance with the system warnings. Drivers with a high school education reported less annoyance with auditory warnings and the brake pulse warning than did more highly educated drivers. The average questionnaire response of drivers within each education level is shown in Figure 28, a lower questionnaire response indicating more annoyance with the warnings. There were no differences in the annoyance of the seat vibration warning by education level.

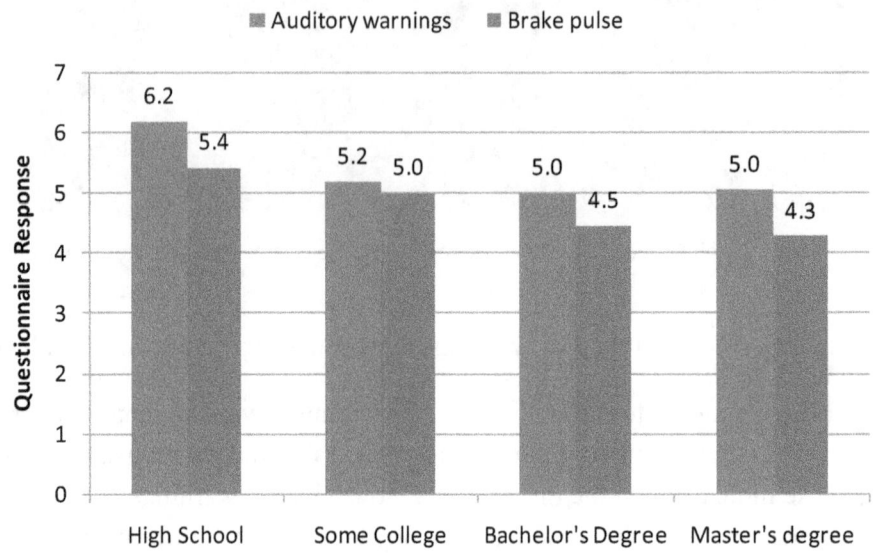

Figure 28. Annoyance with warnings by warning type and education level

3.2.3 Perceived Usefulness

Twelve numerical questionnaire items and two open-ended response questions addressed drivers' perception of the usefulness of the integrated system. The distribution of drivers' percent positive responses to these questions is shown in Figure 29 and Figure 30. Eighty one percent of drivers rated at least half of the 12 usefulness questions positively. Figure 29 shows the distribution broken down by age group. Younger drivers rated the usefulness of the system lower than the other age groups, and older drivers rated usefulness higher than the other age groups. The distribution of drivers' feedback about usefulness of the system is broken down by gender in Figure 30. Female drivers rated the usefulness of the system slightly higher than men.

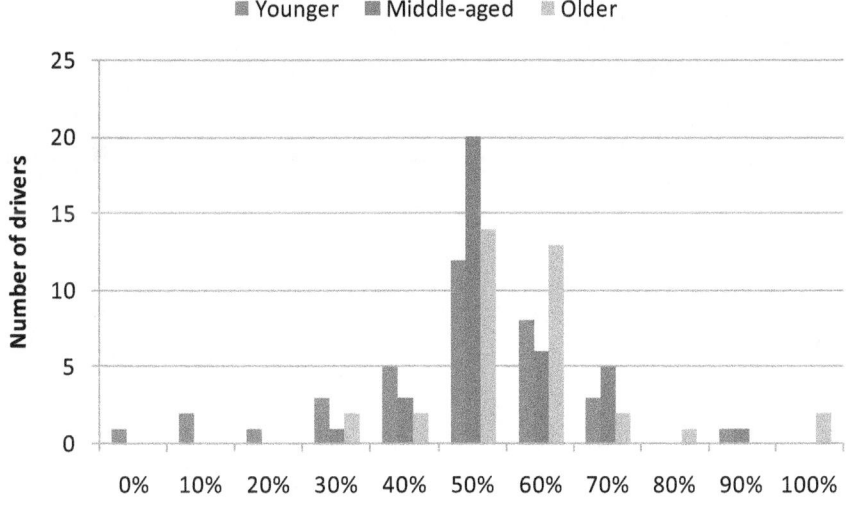

Figure 29. Distribution of perceived usefulness by age group

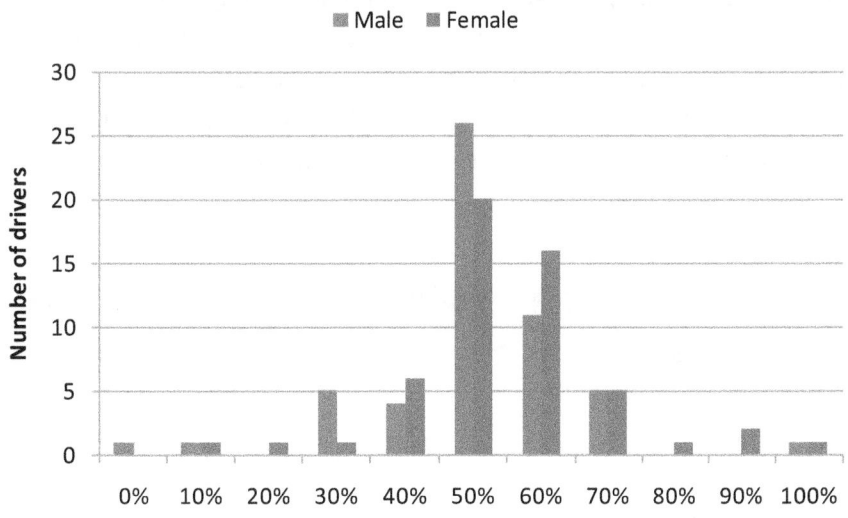

Figure 30. Distribution of perceived usefulness by gender

Figure 31 shows drivers' responses to two questions addressing overall usefulness of the system. Over 80 percent of drivers were satisfied with the system as a whole and found the warnings to be helpful.

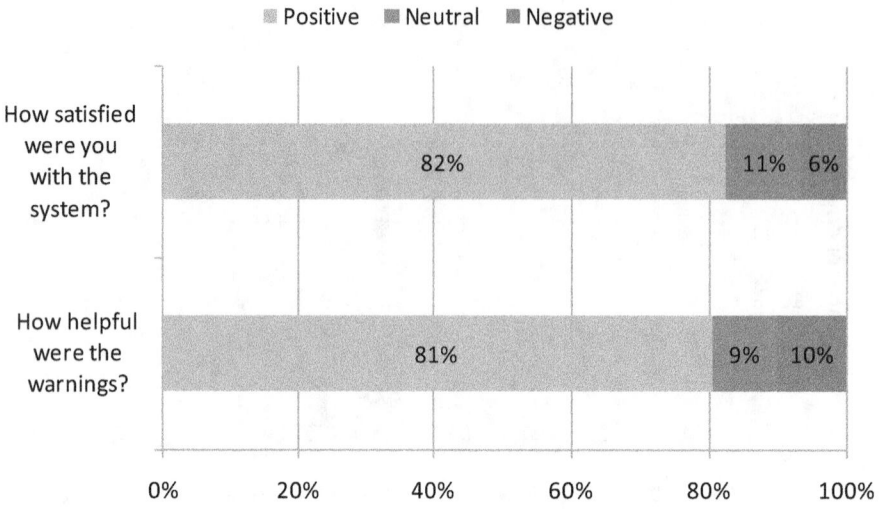

Figure 31. Drivers' opinions about overall usefulness of the integrated system

Drivers' feedback about the safety impact of the system is shown in Figure 32. While 82 percent of older drivers felt the system would increase their safety, only 69 percent of younger drivers agreed. Similarly, all older drivers felt that the system made them more aware of traffic around them and their position of their car on the road while 81 percent of younger drivers agreed. Overall, 90 percent of drivers reported that the system increased their awareness.

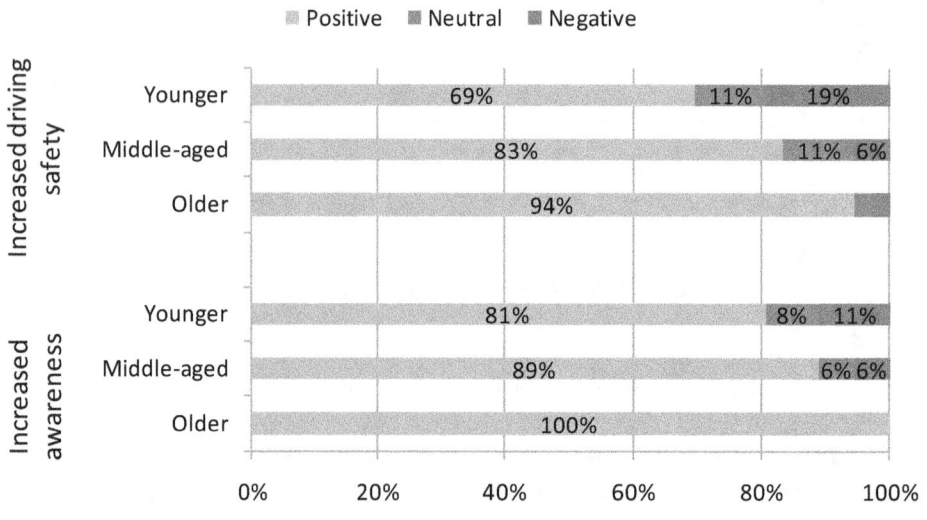

Figure 32. Drivers' opinions about the safety impact of the integrated system

Around 14 of the 31 drivers who attended the focus group reported that the integrated system helped them to avoid a crash. Of these reported near crashes, one was a potential rear-end crash, one involved a sleepy driver who was alerted by a drift warning, and one helped a driver avoid a pedestrian. The other reported incidences involved lane changes and vehicles in blind spots.

Accuracy of the system warnings is related to usefulness of the integrated system. If warnings are not being issued for valid threats, they are not useful to the drivers. The following questions asked drivers about the nuisance alerts they received from the system. In this subjective feedback, the determination of what was considered a nuisance alert was determined by the driver.

Figure 33 illustrates the responses to the question, "the integrated system gave me warnings when I did not need them," broken down by alert type. In these responses, a positive response indicates that the driver did not receive many nuisance warnings. Overall, 63 percent of drivers reported receiving warnings that they did not need, but over half of the drivers indicated that they received warnings with about the right frequency. These results indicate that, while drivers acknowledge the presence of nuisance warnings, they are not particularly bothered by them. Some drivers did comment in the focus group that the presence of nuisance warnings caused distrust in the system or made them begin to ignore the warnings. Within the alert types, the most common nuisance alert type reported by drivers was the side-hazard alert, followed by drift warnings and hazard ahead warnings. Only 31 percent of drivers reported getting a sharp curve warning when they did not need one.

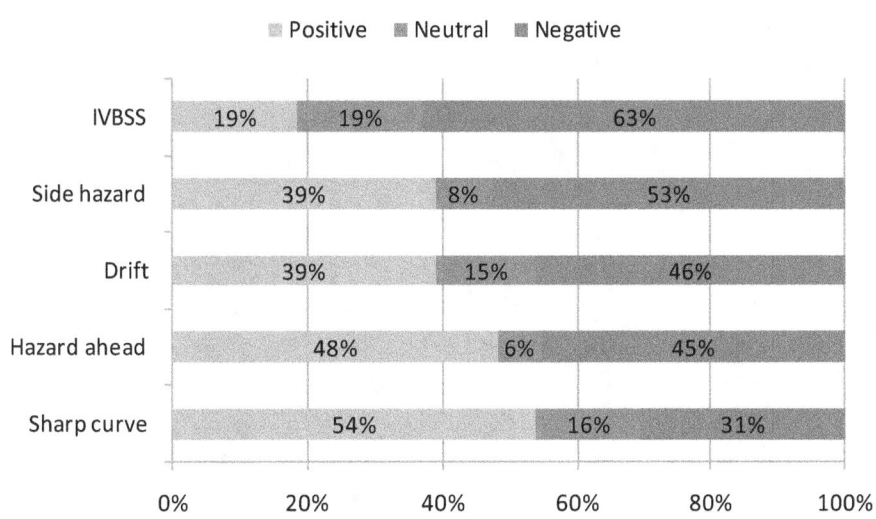

Figure 33. Drivers' reported experience with nuisance warnings

Drivers were also asked about the frequency with which they received nuisance alerts and whether they were annoyed by the nuisance alerts. Figure 34 illustrates the results to both of these questions. There is a noticeable difference in the reported frequency of nuisance alerts among age groups. Sixty-seven percent of older drivers and only 36 percent of younger drivers had positive feedback about the frequency of nuisance alerts. Similarly, 56 percent of younger drivers and only 17 percent of older drivers reported being annoyed by the nuisance alerts.

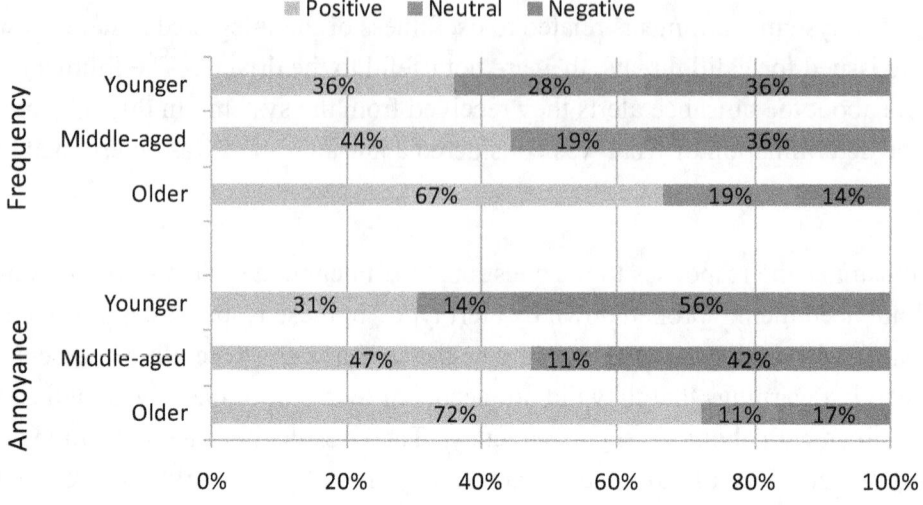

Figure 34. Responses about frequency and annoyance with nuisance alerts

Similar to the effects of age on the perception of nuisance warnings, there was a trend in reported annoyance of nuisance warnings based on number of years of driving experience. Drivers with more experience reported less annoyance with the nuisance warnings than less experienced drivers. Figure 35 shows the average questionnaire responses of drivers with different years of experience, a lower questionnaire response indicating less annoyance with the nuisance warnings.

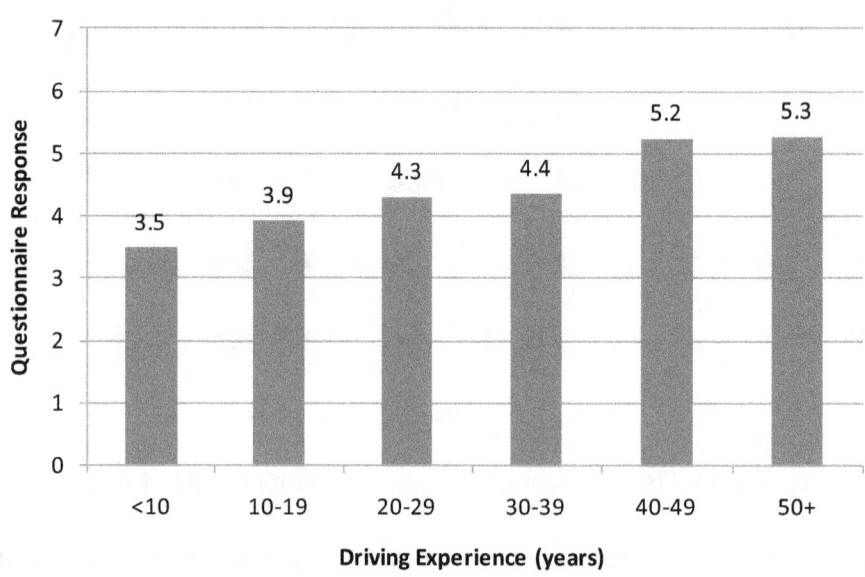

Figure 35. Reported annoyance with nuisance warnings by driver experience

During the focus groups, drivers were asked which of the alert types they would choose to keep if they could only have two or three, rather than the integrated system. Almost all drivers indicated that they preferred the lateral system (side hazards, drift, and blind spot lights) to the longitudinal system (forward-crash warning and curve-speed warnings). Drivers generally found these elements of the system to be the most useful.

3.2.4 Ease of Learning

One question in the focus group sessions addressed ease of learning. The 31 drivers who participated in the focus groups were asked, "after the integrated system was enabled, how long did it take you to become familiar with the system?" Most drivers reported that they were familiar with the system within a few days. None of the drivers reported any problems or challenges with learning how to use the integrated system but a few drivers mentioned that it took a bit longer to become familiar with the warnings that they did not receive frequently.

The equipped vehicle features a text display that shows the type of warning that is being issued. About half of the drivers in the focus group responded that over time they were able to differentiate the different types of warning without looking at the display.

3.2.5 Advocacy

One questionnaire item addressed advocacy for the integrated system. The first asked, "would you like to have the integrated system in your personal vehicle?" The intent of this question was to determine if, regardless of cost, drivers would prefer to drive with the system or without. Seventy-eight drivers said that they would like to have the system in their personal vehicle. A breakdown of responses by age and gender is shown in Figure 36. Younger drivers were slightly less likely to want to drive with the integrated system than drivers in the other age groups.

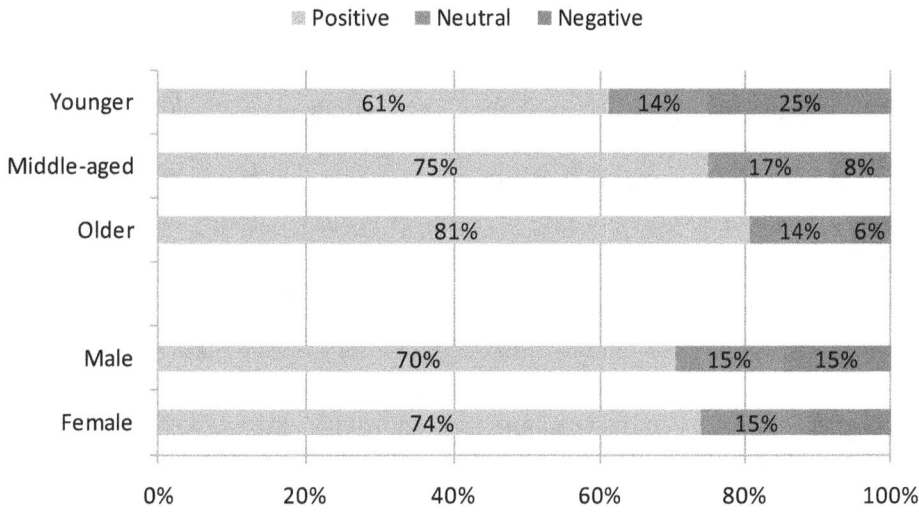

Figure 36. Drivers' willingness to drive with the integrated system

3.2.6 Driving Performance

Two questions were included to assess how the integrated system affected driving performance, with the intent of soliciting feedback on whether or not driving with the system would create any unintended consequences. The results of the first question, "as a result of driving with the integrated system, did you notice any changes in your driving behavior?" indicate that, while 75 percent of drivers felt that the system changed the way they drove, the majority of changes they experienced was positive. The breakdown of results to this questionnaire item is shown in Figure 37. Drivers reported an increase in turn signal use, an increased use of caution and concentration, an increase in awareness of their surroundings and the position on the road, and driving to avoid triggering warnings. Seven drivers reported the potentially negative behavior adaptation; five drivers said they were less likely to check their blind spots when the system was enabled due to the assistance of the blind spot monitors, one driver said that he was hesitant to give pedestrians and cyclists as much room as he generally would so as not to set off a warning, and one driver reported that he increased his frequency of texting while driving. Five of the seven drivers with negative behavior adaptations were male.

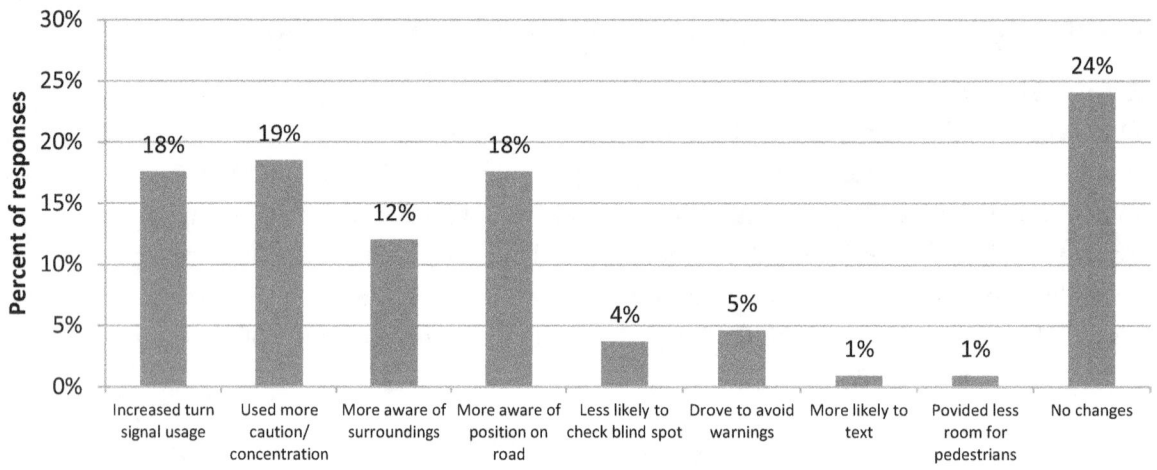

Figure 37. Reported changes in driving behavior due to driving with the integrated system

Thirty-nine percent of drivers responded positively to the second question, "did you rely on the integrated system." The breakdown of these responses is shown below in Figure 38. Three quarters of the drivers who reported a reliance on the system said that they relied on the blind spot monitors rather than turning their head to check their blind spots when changing lanes. Other drivers said they relied on the system to help maintain their alertness and to help them stay in their lane. One driver, a younger male, reported that, "because of [the integrated system] I knew I could pay less attention to the road and text, multitask, eat, make phone calls, etc." Of all elements of the integrated system, the blind spot monitors were the most likely to cause a reliance.

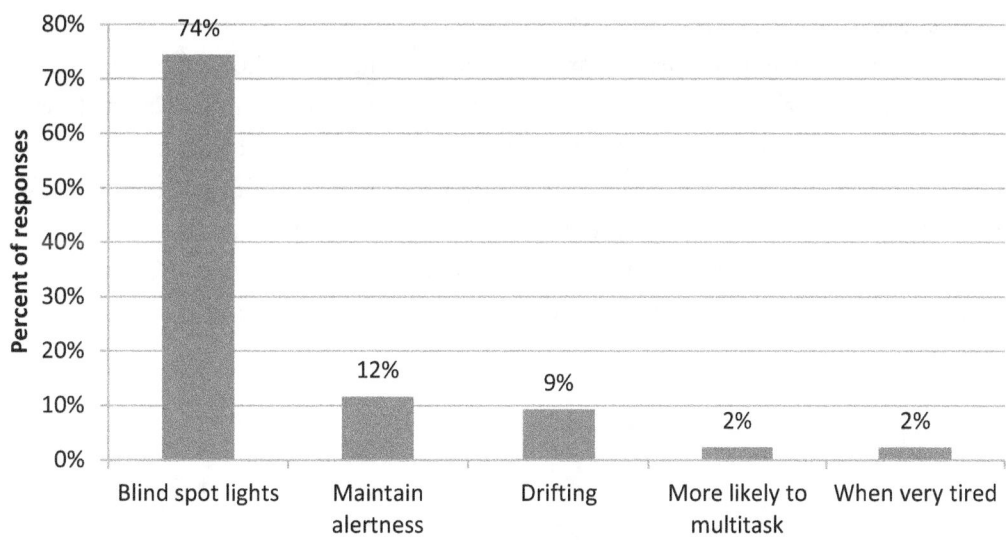

Figure 38. Reported reliance on the integrated system

3.3 Driver Acceptance by Driver Experience Variables

This analysis explores the differences in driver opinion based on their experience in the field test. The list of variables used in this analysis was shown in Table 26. These variables were selected because they quantify elements of system performance that could affect driver acceptance. The number of each type of alert the driver receives can have an impact on their familiarity with the warnings and similarly, the rate at which they experienced warnings can alter their mental model of the warnings. The rate of alerts that drivers received that were not triggered for a valid threat gives insight into how the driver views the appropriateness of the warnings. Driving patterns can affect the way the system performs due to traffic and road geometry. Finally, the frequency with which drivers were in conflict scenarios can give insight into the frequency with which they potentially received alerts that helped them to avoid collisions.

The results discussed in this section are statistically significant based on the results of a paired t-test that compares the numerical questionnaire responses of two groups of drivers; the 36 drivers with the lowest values for each variable, and the 36 drivers with the highest values for each variable. For example, for alert numbers, the questionnaire responses of the 36 drivers with the fewest alerts were compared to the responses of the 36 drivers who received the most alerts. The questionnaire responses of the remaining 36 drivers were not included in the analysis. This method was selected to compare drivers with a larger degree of difference in their experience with various elements of the system. In a situation where drivers did not respond to a questionnaire item (for example, if a driver did not receive any FCW alerts, he/she would not respond to questions about the FCW warnings), they were not included in the analysis. Significance for these results was set at $p < 0.05$. Error bars represent the 95 percent confidence interval.

3.3.1 Alert Rate

The alert rate or number of alerts per 100 miles of driving indicates the frequency with which divers received alerts during the field test. The questionnaire items that showed a statistically significant difference for drivers with different overall alert rates are shown in Figure 39. Drivers who experienced a higher alert rate reported better understanding of the warnings. These results may indicate that if drivers receive more warnings they may build a better mental model of how the system works as they become more familiar with the warnings. Similar results were found for the total number of alerts drivers received during the field test.

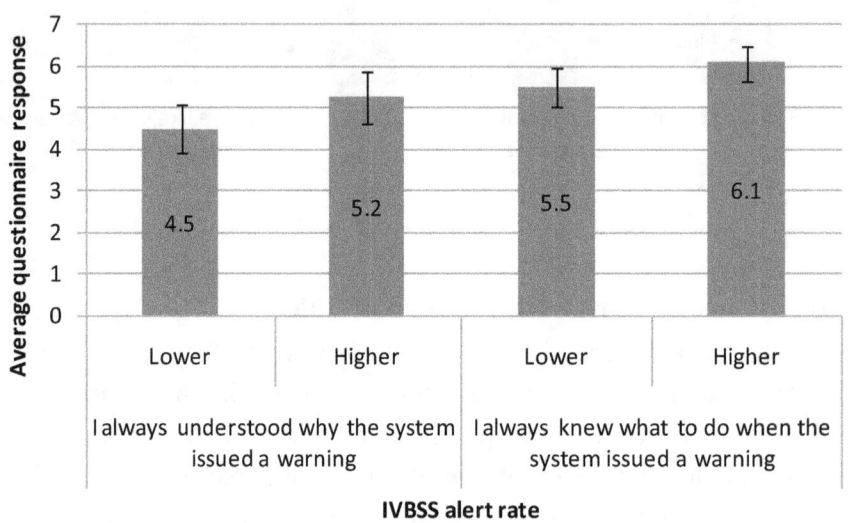

Figure 39. Questionnaire responses by overall alert rate

3.3.2 Driving Patterns

Where and when a driver travels can affect how the system behaves with respect to the frequency and appropriateness of alerts. This section breaks down the questionnaire responses by driver's average trip length, breakdown of daytime/nighttime driving, and freeway/non-freeway driving. Total mileage during the treatment was also explored, but no questionnaire responses produced significant results.

A driver's average trip length represents a driver's driving pattern. A shorter trip length suggests that drivers stay close to home and make many shorter trips, likely on surface streets. A longer trip length suggests that a driver frequently travels further away and is in the car for longer periods of time. One questionnaire response showed significance by trip length; drivers who made shorter trips agreed more strongly with the statement, "the integrated system made driving easier" than the drivers who more frequently made longer trips. These results are illustrated in Figure 40 where a higher questionnaire response indicates stronger agreement that the system makes driving easier.

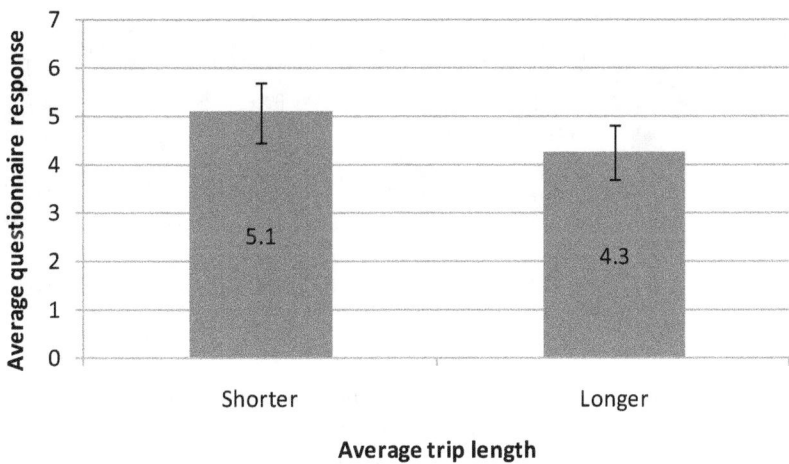

Figure 40. Drivers' responses to the statement, "the integrated system made driving easier"

Two questionnaire items showed a difference in response based on a driver's proportion of daytime and nighttime driving. Drivers who had the highest proportion of nighttime driving reported lower distraction and better understanding of the system warning.

Three questionnaire items showed a significant difference based on the road type more frequently traveled by the driver. Freeway driving and non-freeway driving can produce a different response from the system due to the presence of, and the nature of the surrounding traffic, and can also change a driver's perception of the usefulness of the warnings due to their ability to predict what the surrounding traffic will do. There are fewer unexpected maneuvers by surrounding vehicles on the freeway than on surface streets, where vehicles are entering, exiting, and cutting across.

The results of the questionnaire items broken down by drivers who drove a lower and higher proportion of their mileage on the freeway are shown in Figure 41. Drivers with the highest proportion of their mileage on non-freeway roads agreed more strongly that the system was predictable and consistent. Similarly, this group of drivers reported better understanding of why the system issued warnings. Finally, this group of drivers agreed more strongly that the system issued warnings when they did not need them (a higher questionnaire score indicates more nuisance warnings). These results suggest that the system is more predictable and issues more accurate and useful warnings on surface streets than on freeways.

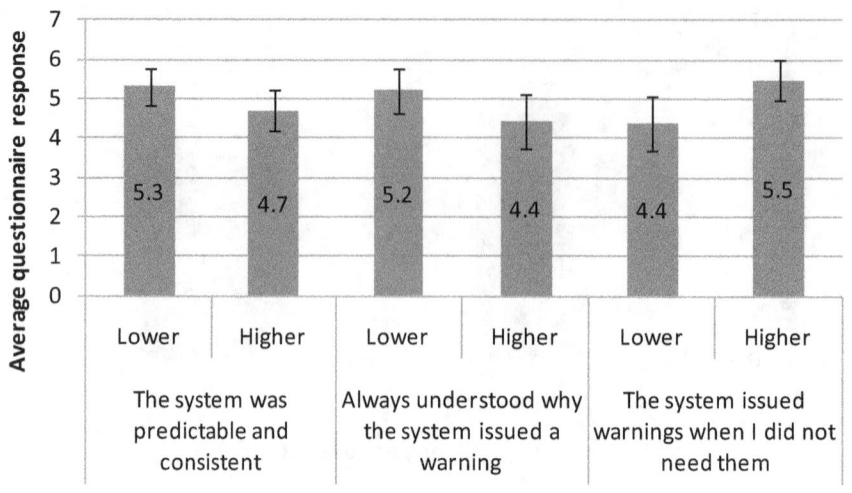

Figure 41. Questionnaire responses broken down by drivers' proportion of freeway driving

3.3.3 Conflict Rates

Drivers with both higher overall conflict rates and higher rear-end conflict rates had more agreement with the statement, "I always understood why the integrated system provided me with a warning" (higher response indicates more reported understanding of the warnings). It is possible that drivers who were in conflict scenarios more frequently received a higher proportion of valid warnings. Valid warnings have a clear, obvious threat that can be more easily understood by drivers. These results are illustrated in Figure 42.

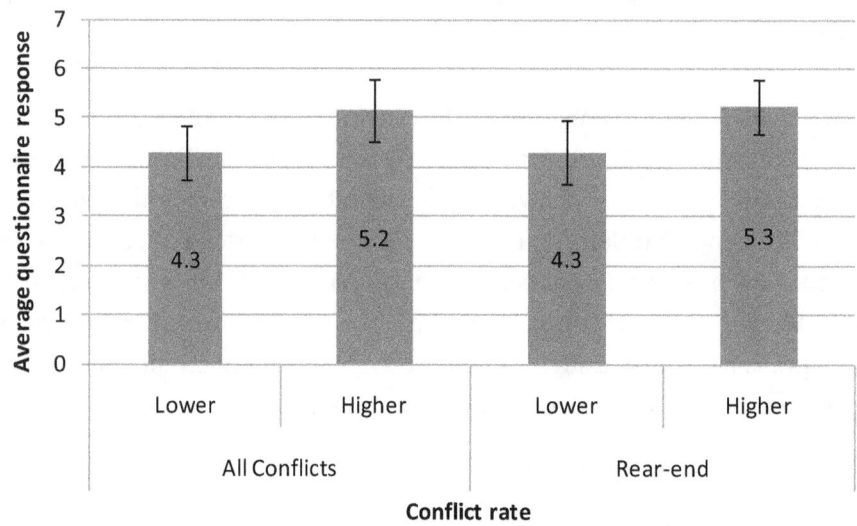

Figure 42. Drivers' responses to the questionnaire item, "I always understood why the system provided me with a warning"

3.3.4 False Alarm Rates

False alarm rates refer to the frequency with which the system issues warnings when no valid threat is present. A driver's frequency of exposure to false alerts can affect both their trust in a system, as well as their annoyance with the warnings. For forward-collision warnings, a false alert is defined as an alert that is issued when there is not an in-path target. A false curve-speed warning is defined a warning that is issued when a driver does not traverse a curve. Lane-change/merge warnings and imminent drift warnings (also referred to as side-hazard warnings) are considered false if they are issued when an adjacent side target is not present, and drift warnings are considered false if an alert is issued when the driver does not cross a lane boundary. Validity of the warnings was determined through video analysis of the warning scenarios.

For each alert type, drivers' reported exposure to false alerts was consistent with their actual exposure to false alerts. Figure 43 shows average questionnaire responses to the statement, "The system gave me warnings when I did not need them," for drivers who received the least and most false alerts for each alert type. Each questionnaire item shown below was specific to alert type (for example, "the system gave me forward-collision alerts when I did not need them," "the system gave me sharp curve warnings when I did not need them," etc.), and a higher value represents stronger agreement with the statement. The results for side-hazard alerts only include the rates of false lane-change/merge warnings, as very few drivers received false imminent drift warnings. Results for each alert type were statistically significant, indicating that drivers had an accurate perception of the false alerts that they received.

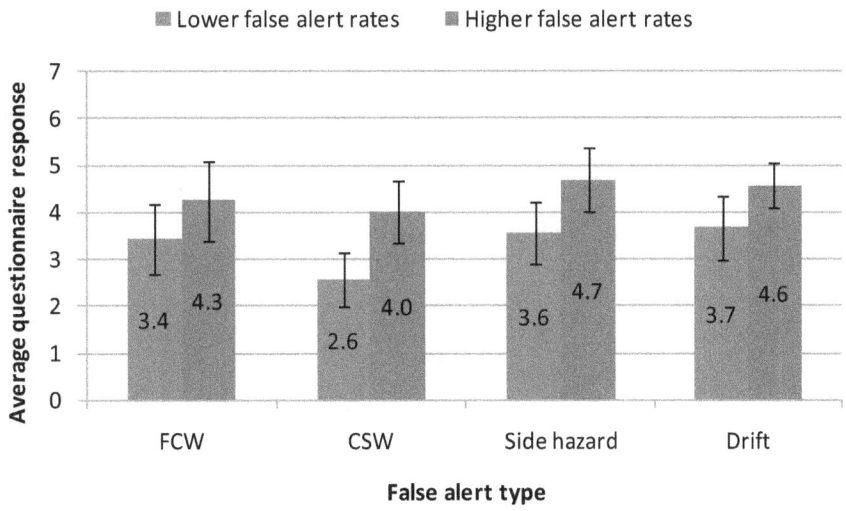

Figure 43. Responses to questionnaire items relating to drivers' exposure to nuisance alerts

One alert type showed a significant difference in drivers' overall perception of the frequency of nuisance alerts. Drivers who received a high rate of false cautionary drift warnings were more likely to say that they received nuisance warnings too frequently. These results are illustrated in

Figure 44, where a lower questionnaire response indicates more nuisance warnings. Due to the overall frequency of LDW-C warnings, the rate of false LDW-C alerts was much higher than the rates of other types of false alerts. Drivers who received many LDW-C alerts would perceive a higher number of false warnings overall.

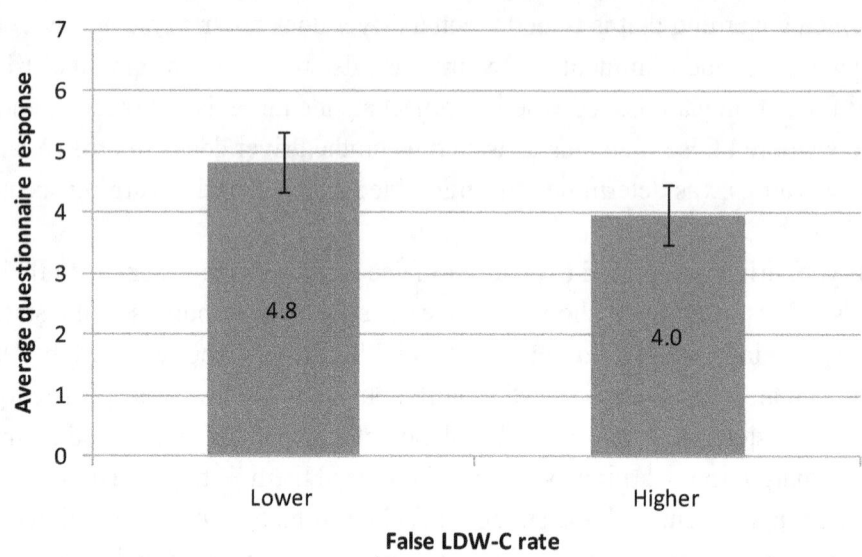

Figure 44. Responses to the statement, "Overall I received nuisance warnings (1=Too frequently, 7= Never)" by rate of false LDW-C warnings

Figure 45 presents statistically significant differences between drivers with the lowest and highest rates of false alerts. These questions pertain to both the presence and frequency of nuisance warnings. Drivers who received more false warnings had stronger agreement that the system issued nuisance warnings and also reported that the system issued nuisance warnings too frequently. These results, along with the results presented earlier in the section, indicate that drivers can accurately detect the presence of nuisance warnings.

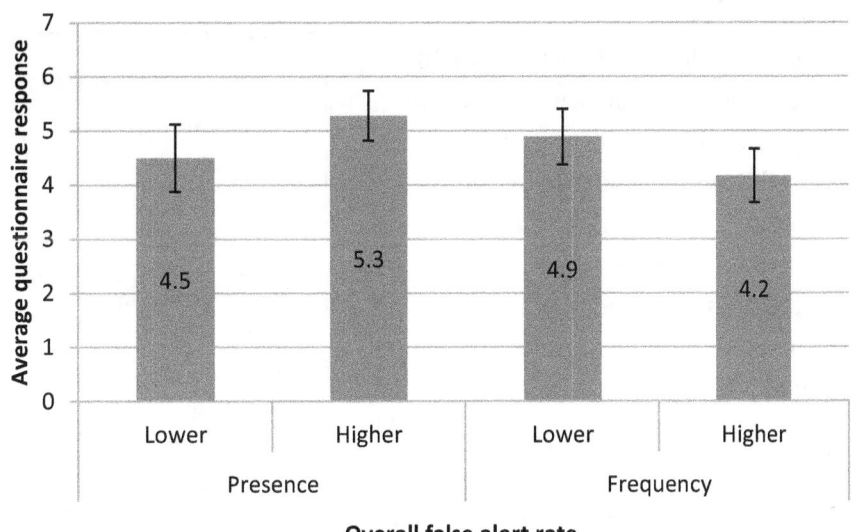

Figure 45. Responses to the statements, "the system gave me warnings when I did not need them" and "overall, I received nuisance warnings… (1= Too frequently, 7= Never)" by overall false alert rate

4. System Capability

This section provides results of the system capability analysis that was conducted for the sensors, warning logic, driver-vehicle interface, and robustness of the integrated system. The performance of the sensors was evaluated in terms of their ability to accurately determine the presence of a threat. The warning logic was examined in terms of the system's decision-making to alert drivers to driving conflicts that might lead to rear-end, lane-change/merge, or road-departure crashes. The driver-vehicle interface was evaluated in terms of its capability to properly convey visual, audible, and haptic information to the driver and system controls were assessed in terms of frequency of use. System robustness was appraised by its availability during the field test.

> **HIGHLIGHTS**
> - Overall, alerts had a very high degree of accuracy.
> - Alerts issued for forward stationary targets were issued mostly for out-of-path targets, indicating a low degree of accuracy for this type of FCW warning.
> - Drivers responded to forward threats more quickly and more assertively when they received FCW alerts.
> - Drivers exhibited higher deceleration levels when approaching curves with the system enabled.
> - Drivers resumed their lane position after drifting more assertively when the system was enabled.
> - Drivers had fewer drift warnings with the system enabled, indicating better lane-keeping behavior.
> - With the system enabled, drivers showed a 46 percent reduction in drifts to the left, the type of drift that can lead to a head-on collision.

4.1 Sensors

This analysis was based on a sample of 16,915 alert videos which were reviewed to characterize the performance of the forward-looking, side-looking, and lane-tracking sensors of the integrated safety system and the system's ability to determine upcoming road geometry. A detailed breakdown of the alerts analyzed for this analysis is located in Appendix C, and definitions of each coded variable discussed in this section are located in Appendix D.

4.1.1 Forward-Looking Sensors

Evaluation of forward-looking sensor performance was based on the analysis of 851 FCW alerts in which alerts were characterized by target location. Target location refers to whether the detected object was in the equipped vehicle's intended lane of travel at the time of the alert. The system was designed to issue alerts for in-path objects only. Roadside signs, overhead bridges, guard rails, and vehicles in adjacent lanes were all considered out-of-path targets. Performance was measured by the proportion of FCW alerts issued for out-of-path targets, and is broken down by moving and stopped targets. The distribution of alerts issued for out-of-path targets are provided later in this section by target type, speed bin, and vehicle location (as defined in the video coding manual of the MDAT in Appendix D).

Figure 46 shows the proportion of FCW alerts that were issued in response to in-path targets broken down by target type. Overall, half of the alerts were issued for in-path targets. Alerts issued for moving targets had a much higher degree of accuracy (88 percent) than those issued for stopped objects (eight percent).

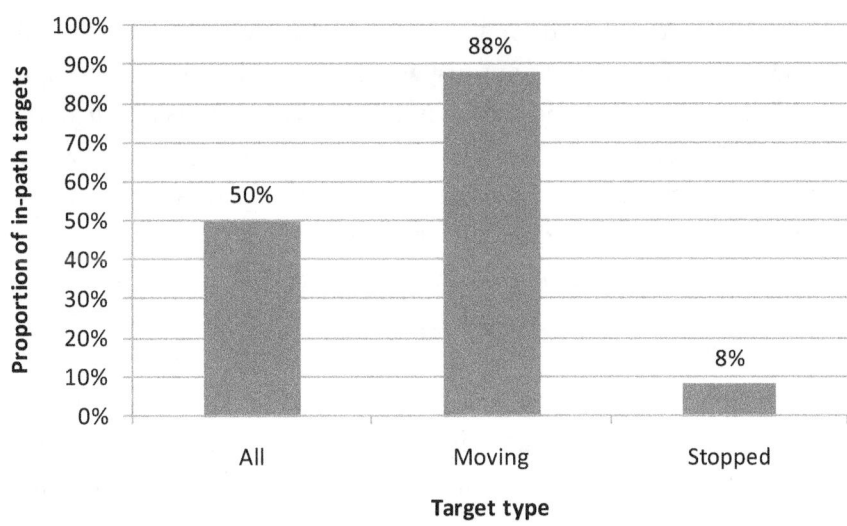

Figure 46. Proportion of in-path targets for FCW alerts by target type

The proportion of moving and stopped in-path targets is further broken down by speed bin in Figure 47. The highest proportion of stopped in-path targets was in the lowest speed bin, where it is most likely that a driver would encounter an in-path stopped vehicle or a cross-path moving vehicle (because cross-path vehicles are moving in a direction perpendicular to the direction of the host vehicle rather than towards or away, the system registers them as stopped objects). No stopped targets were in-path in the 55 mph and over speed bin, the speed bin that frequently represents freeway driving. The proportion of moving targets that were in-path was very high in the lowest three speed bins, but dropped considerably when drivers were traveling on freeways. A closer investigation of the 33 out-of-path alerts issued in this speed bin revealed that 30 of the 33 alerts were issued when the vehicle was adjacent to and passing a large truck.

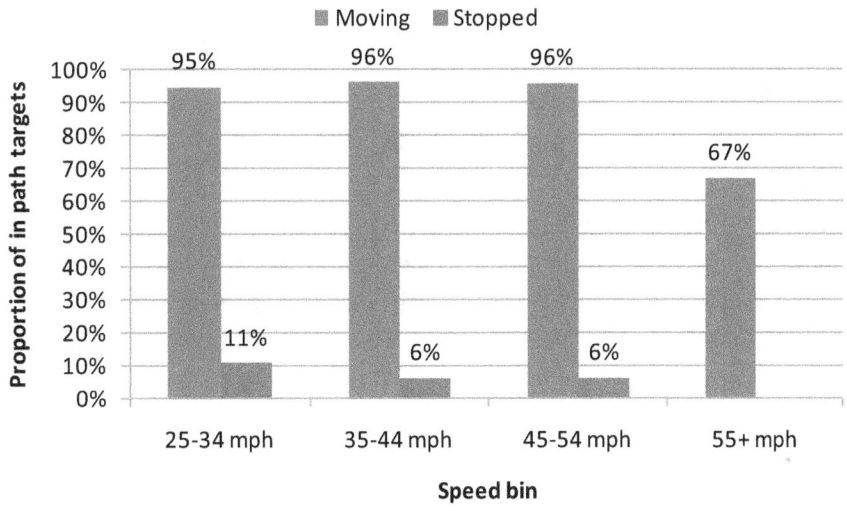

Figure 47. Proportion of in-path targets for FCW alerts by target type and speed bin

71

The target type of the 355 alerts issued for stopped, out-of-path targets is broken down by speed bin in Figure 48. Overall, 78 percent of out-of-path stopped object alerts were issued for roadside signs or objects. It is common for the integrated system to issue an alert for a stopped object that is ahead as the vehicle is entering a curve. While these objects are directly in-path at the time the alert is issued, the driver perceives them as out-of-path objects because they are not in the roadway. While results for the three lower speed bins are similar, out-of-path alerts issued at higher speeds are less frequently issued for roadside signs and objects, and more likely issued for overhead bridges and signs. These results are most likely the result of differences in road geometry. It is important to note that out-of-path stopped object alerts are very rare. Of the 355 alerts for out-of-path stopped objects, only eight were issued in the over-55 speed bin. Only five alerts were issued for overhead bridges and signs during the field test, one in the 35-44 mph speed bin, one in the 45-54 mph speed bin, and three in the 55 and over speed bin.

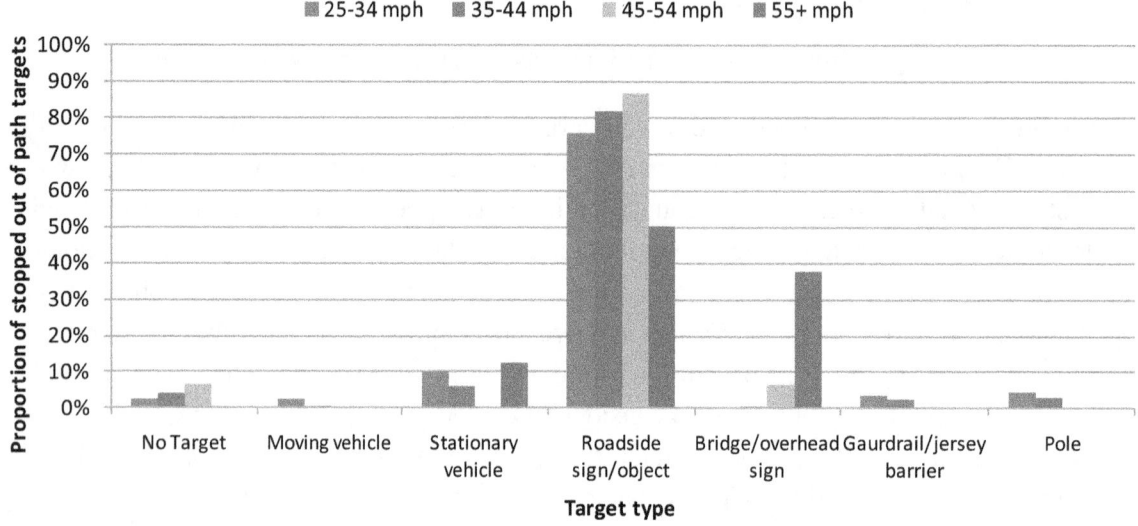

Figure 48. Target type of alerts issued for stopped out-of-path targets, by speed bin

Alerts issued for stopped objects are rarely in-path of the host vehicle. The probability that an alert issued for a stopped object is out-of-path, based on the analysis of 386 stopped-object alerts is shown in Figure 49. These results indicate that there is a higher probability that an alert issued for a stopped object is out-of-path when the vehicle is at some point in a curve than when they are on a straight road. As mentioned earlier, it is common for the system to issue alerts for roadside objects when the vehicle is entering, exiting, or traversing a curve.

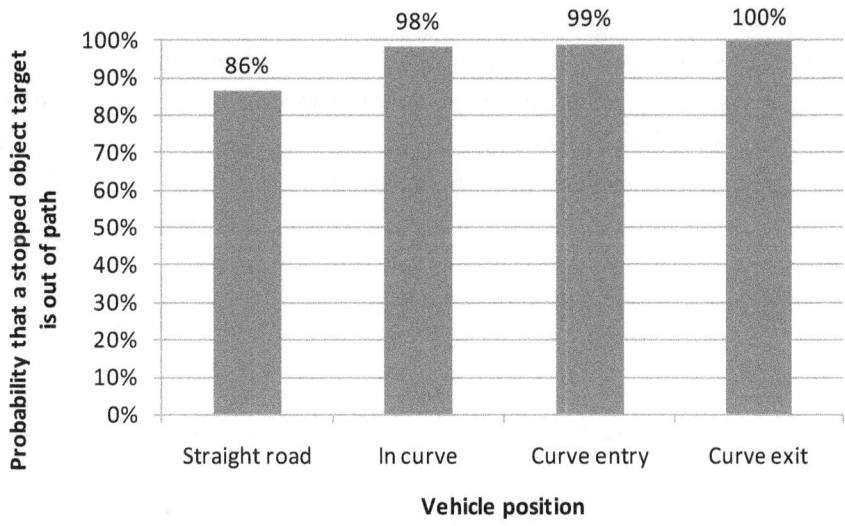

Figure 49. Probability of out-of-path stationary object alert, by vehicle position

The straight road condition shown above in Figure 49 represents a total of 183 FCW alerts issued for out-of-path stationary objects on straight roads. These alerts were further broken down by range to determine the ranges at which the system more commonly issues alerts for out-of-path stationary targets on straight roads. Range refers to the distance to the target at the time the alert is issued. Figure 50 illustrates that the majority of these alerts (77 percent) is issued for targets over 30 meters away from the host vehicle.

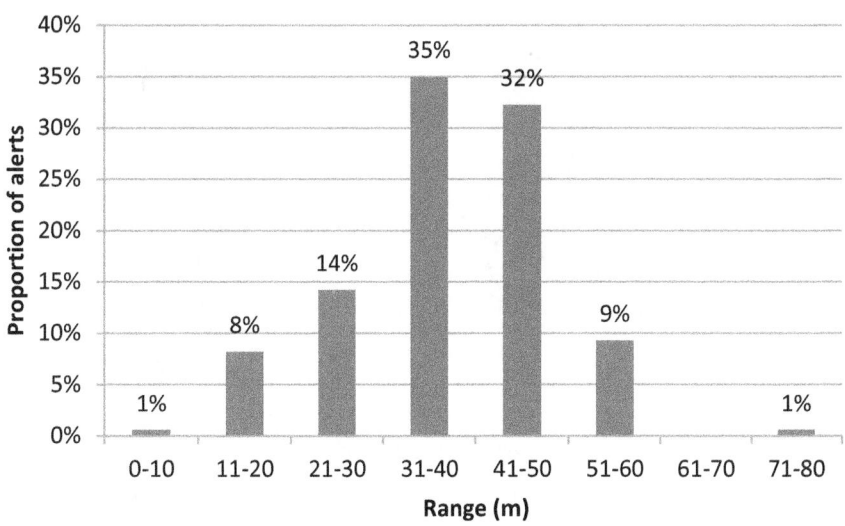

Figure 50. Range of stationary targets triggering out-of-path FCW alerts on straight roads

4.1.2 GPS and Map data

Performance of the GPS and map system was based on the analysis of 919 CSW warnings. The purpose of the CSW system is to alert drivers if they are approaching a curve too quickly so they can decelerate to a safe speed before entering the curve. The system uses the vehicle's speed as well as GPS and a map system to determine the upcoming curvature of the road. The capability of the map system is characterized in terms of the proportion of CSW alerts that were issued when the driver traversed a curve. Alerts issued when the driver does not traverse a curve are not helpful to the driver and are considered to be false.

Overall, 78 percent of CSW alerts were issued when the host vehicle traversed a curve. The breakdown of these alerts by road type is shown in Figure 51, which shows that CSW alerts were slightly more accurate on arterial roads than on freeways. Also shown in Figure 51 is the proportion of alerts that were inaccurately triggered by different scenarios. When a vehicle passes a road split (such as a freeway exit ramp), the system's GPS may interpret that the vehicle is on a trajectory to traverse the curve when the driver intends to continue straight on the freeway. Similarly, the system may interpret a perpendicular roadway (either an intersection or an overpass) or adjacent road as a turn the vehicle intends to take. Twenty-two curve-speed alerts were issued due to intersections and overpasses during the field test.

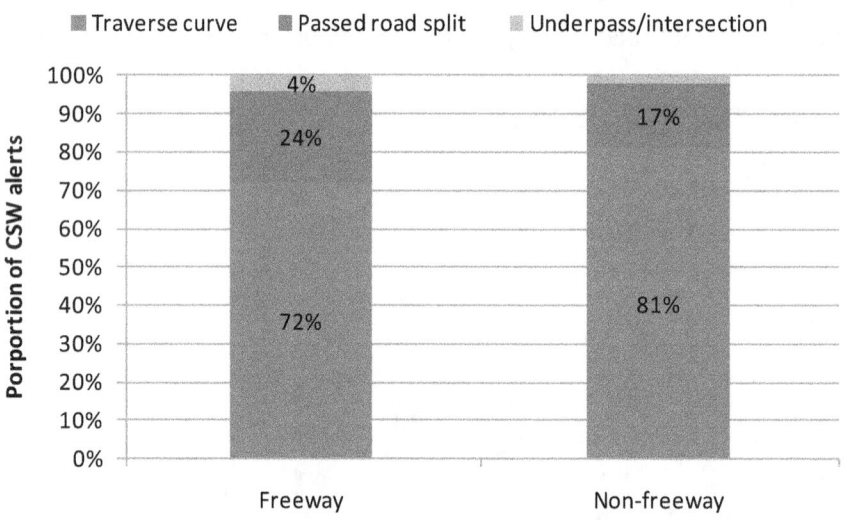

Figure 51. Curve-speed alerts by road type

4.1.3 Side-Looking Sensors

Performance of the side-looking sensors of the integrated system was characterized in terms of the presence and relative location of side targets triggering LCM and LDW-I alerts. This analysis is based on the review of 1,336 LCM alerts and 2,501 LDW-I alerts. The aim of the lateral warning systems is to detect adjacent targets that pose a direct threat during a lane-change maneuver or a drift out of the driver's lane. Adjacent targets are defined as targets occupying the space adjacent to the vehicle (another travel lane, shoulder, or roadside), which are either beside the host vehicle or in the closing zone (in close proximity behind the host vehicle). Targets in the adjacent lane ahead of the vehicle do not pose a threat because they allow the driver enough room to safely make a lane change, and targets two or more lanes over do not pose a threat to the vehicle during a lane change. The main performance measure is the proportion of LCM and LDW-I alerts issued for adjacent targets. For LCM, the added breakdown of target type for adjacent targets is included, as the system is designed to only issue alerts for vehicle targets. The distribution of non-adjacent target alerts is provided by target type.

Ninety-three percent of all side-imminent alerts were issued for adjacent targets. The breakdown of target location for lane-change and imminent-drift warnings is illustrated in Figure 52 and Figure 53, respectively. Eighty-six percent of all LCM alerts and 95 percent of all LDW-I alerts were issued for adjacent targets. The breakdown of 161 LCM and 95 LDW-I alerts issued for non-adjacent targets is shown in Figure 54; of the 165 moving vehicle targets represented, half were issued for vehicles two or more lanes over and half were issued for vehicles in the forward view.

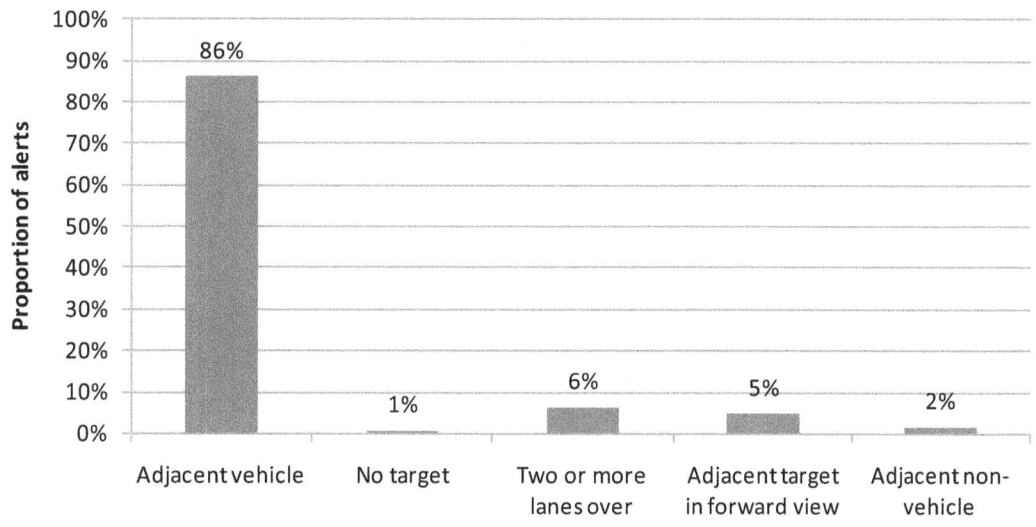

Figure 52. Target location of LCM alerts

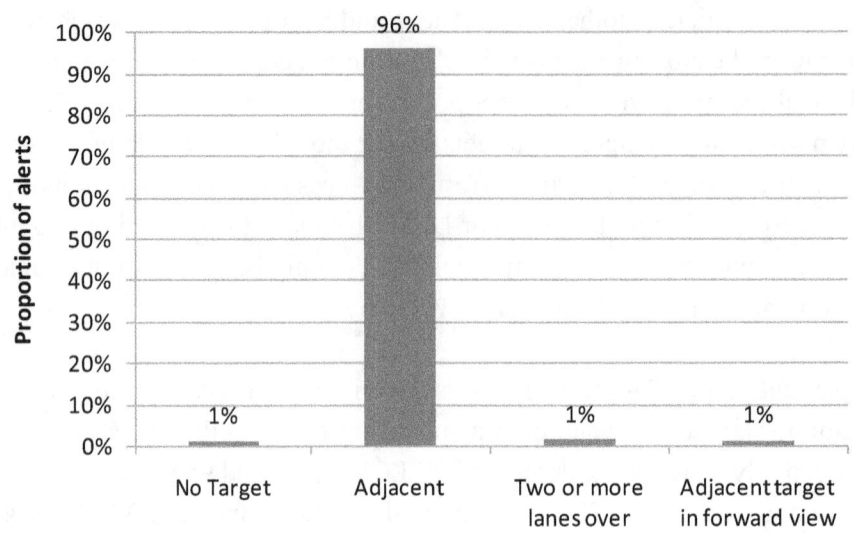

Figure 53. Target location of LDW-I alerts

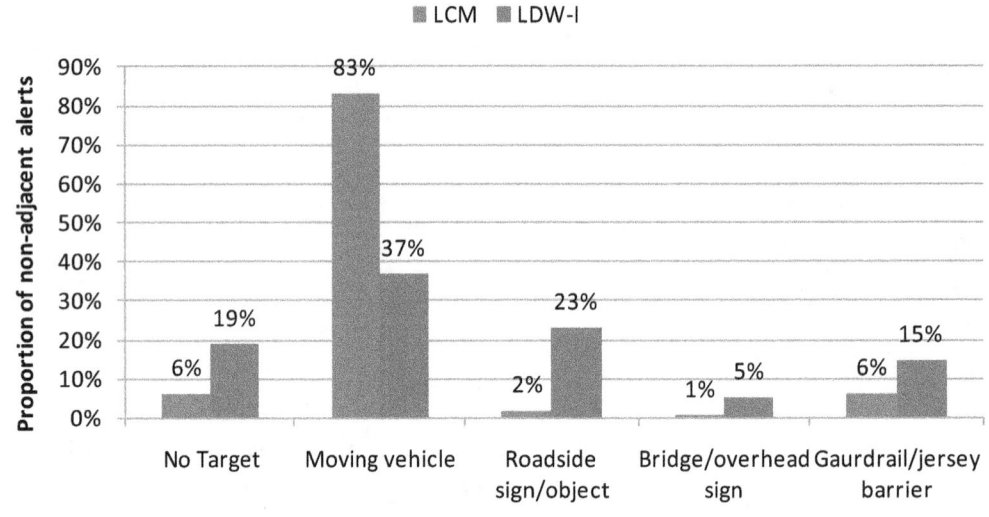

Figure 54. Target type of side-imminent alerts issued for non-adjacent targets

4.1.4 Lane Tracking

Performance of the vision-based lane-tracking sensor used for the LDW function was analyzed by its ability to detect unsignaled lane excursions. LDW-C warnings alert drivers when they are drifting out of their travel lane and no target is present. A sample of 11,308 LDW-C alert videos was analyzed to determine alert validity. This analysis distinguishes between unintentional and intentional excursions; unintentional excursions include alerts issued because the car drifted over the lane line while intentional excursions account for alerts that are issued when drivers intentionally leave their lane of travel (e.g., make an un-signaled lane change or maneuver around an obstacle). LDW-C alerts issued for intentional and unintentional excursions are considered to be valid.

Eighty-six percent of the LDW-C alerts analyzed were valid. These alerts are made up of the intentional and unintentional excursions shown in Figure 55. Invalid alerts were present in a higher proportion of non-freeway alerts than alerts issued on freeways. Freeways generally have more consistent and visible lane markings than non-freeway roads, allowing the lane tracking system to produce more accurate results. Intentional lane excursions were more common on freeways. The proportion of LDW-C alerts issued for intentional lane excursions dropped from 51 percent in the baseline period to 38 percent in the treatment period due to the decrease in unsignaled lane changes discussed in Section 2.

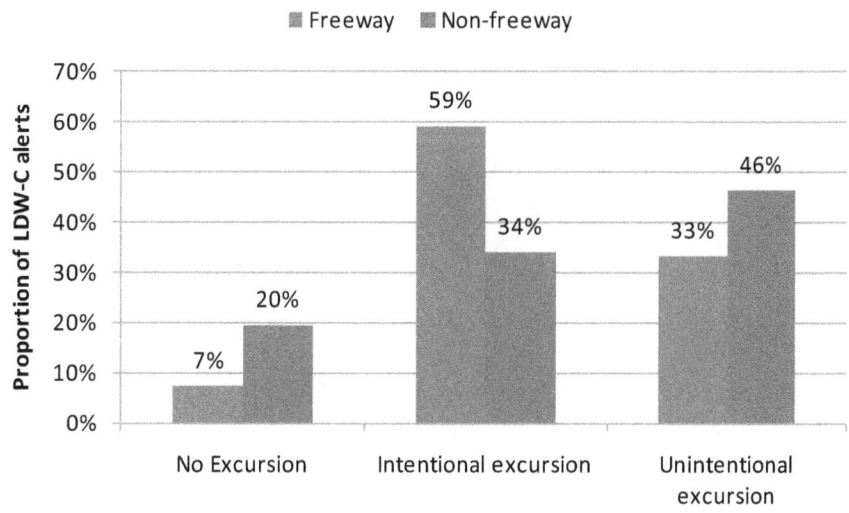

Figure 55. Lane excursion scenario of LDW-C alerts by road type

External factors such as road surface condition, lighting, and weather can affect the accuracy of the vision-based lane-tracking system by obscuring or distorting lane markings. Figure 56 shows the probability that an LDW-C alert is issued when no lane excursion occurred by road surface condition (from video analysis), lighting conditions (based on the lane tracking sensors), and weather conditions (based on the vehicle's windshield wipers). This figure illustrates that while time of day and weather conditions do not show a large effect on the probability of the driver receiving a false alarm, different road surface conditions can have a large effect on the accuracy of the LDW-C alerts. Snow or moisture on the road surface will appear to the lane-tracking camera like a lane line, causing the camera to track incorrectly. For example, a stripe of snow in the center of the roadway between tire tracks can sometimes appear as a lane marking, particularly if the actual lane marking is obstructed by snow or water. Roadways with snow on the surface had the highest probability of creating an inaccurate drift warning, followed by salt residue and standing water. Adverse weather did not have a large effect on the probability of an inaccurate drift warning because lane tracking inaccuracies are caused by standing water or snow rather than precipitation.

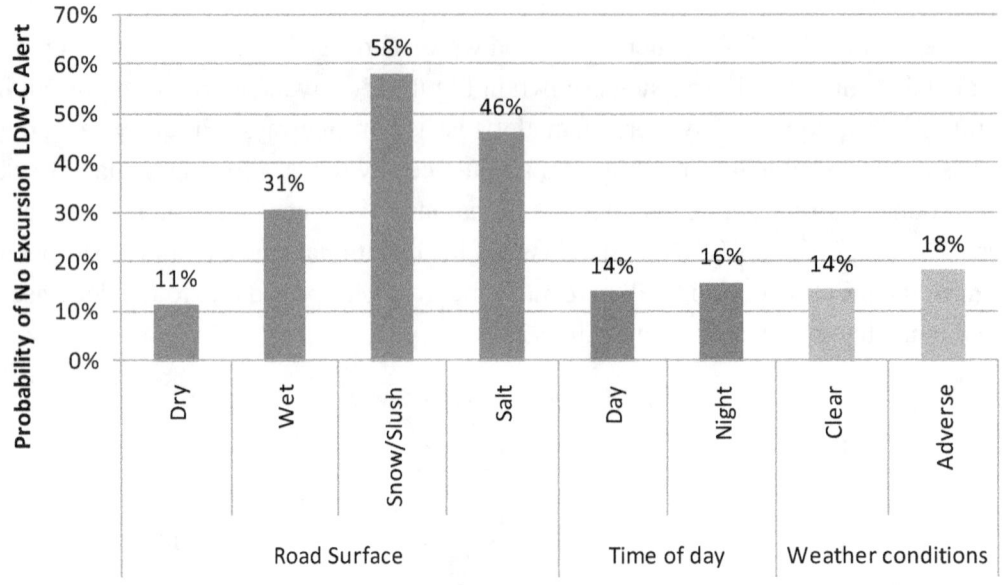

Figure 56. No excursion LDW-C probability by environmental factors

4.2 Warning Logic

The performance of the warning logic was analyzed using a sample of alert episodes where an obstacle or road hazard was in the intended travel path of the light vehicle. Specifically, this analysis focuses on the in-path vehicle or obstacle category shown in Table 27. The "no hazard" and "out-of-path hazard" categories were addressed in the performance of sensors discussed above. A false alert was a warning caused by noise or interference when there is no object or threat present. Out-of-path nuisance alerts were caused by vehicles and objects that are not in the intended path of the host vehicle. In-path nuisance alerts referred to warnings for vehicles that were in the intended path of the vehicle, but were at a distance or moving at a speed that drivers did not perceive as threatening. For instance, forward-crash warnings were issued for lead vehicles turning at intersections. Some of these alerts could be issued based on the system design, but drivers usually perceived them to be unnecessary. In this section, the in-path vehicle or obstacle alerts were analyzed by hazard propensity and driver response.

Table 27. Analysis of system alerts

	No Hazard	In-Path Vehicle/Obstacle		Out-Of-Path Hazard
		Situation Requiring an Alert	Situation Not Requiring an Alert	
Alert Issued	False alert	Appropriate alert	In-path nuisance alert	Out-of-path nuisance alert
No Alert Issued	Appropriate non-alert	Missed alert	Appropriate non-alert	Appropriate non-alert

Performance results of the warning logic are provided based on the analyses conducted on the hazard propensity and driver response to system alerts issued for valid crash threats.

4.2.1 Hazard Propensity

The efficacy of the warning logic to issue appropriate alerts is assessed through the mapping of system alerts to driving conflicts and near crashes. Alerts that correspond with conflicts or near crashes are likely to be considered useful by the driver. Correspondence was determined by the overlap of the conflict duration over a time window ranging from 10 seconds before to 15 seconds after the onset of the alert. In contrast, mismatches between alerts and conflicts or near crashes indicate that the system may have missed a hazardous situation or issued a nuisance alert.

It should be noted that the warning logic designed for the system does not necessarily match the definition of driving conflicts and near crashes in this report. Additionally, many conflicts and near crashes were not alerted because the system suppressed the alerts for a variety of reasons including the use of the turn signals, brake pedal press, lane-change maneuver, occurrence of a prior system alert, and travel speed under 25 mph. Driving conflicts, which are based on driver response, can still occur in these circumstances so is not expected that an alert would be issued in every instance where a conflict or near crash occurred.

Figure 57 illustrates the mapping of valid alerts to driving conflicts. The first set of data in this figure represents the percent of each conflict type for which alerts were issued. Overall, alerts were issued for 11 percent of conflicts. Rear-end conflicts had the lowest rate of alerts due to travel speed and alert suppression due to brake pedal activation. Seventy-one percent of all road-departure conflicts occurred below 25 mph, the minimum operable speed for the integrated system. Additionally, although drivers may be braking for a decelerating lead vehicle and therefore suppressing an alert, they may initially not brake hard enough, requiring hard braking as they get closer to the lead vehicle. Thirty-five percent of road-departure conflicts triggered system alerts, making road-departure conflicts the most consistent with system operation.

The second field of data in Figure 57 represents the proportion of all alerts that were triggered by conflict scenarios, scenarios where drivers reacted strongly. In this figure, FCW alerts map to rear-end conflicts, CSW warnings to curve speed, LCM to lane-change/merge, and LDW to road departure. Overall, 13 percent of alerts were issued when the scenario required a strong driver response. The highest proportion by conflict type was curve speed where 36 percent of alerts were issued. Drift alerts had the lowest proportion of alerts issued for conflict scenarios, likely because drift alerts were very common and most often cautionary warnings. Since these warnings do not indicate an imminent threat, they do not require a strong evasive maneuver by the driver.

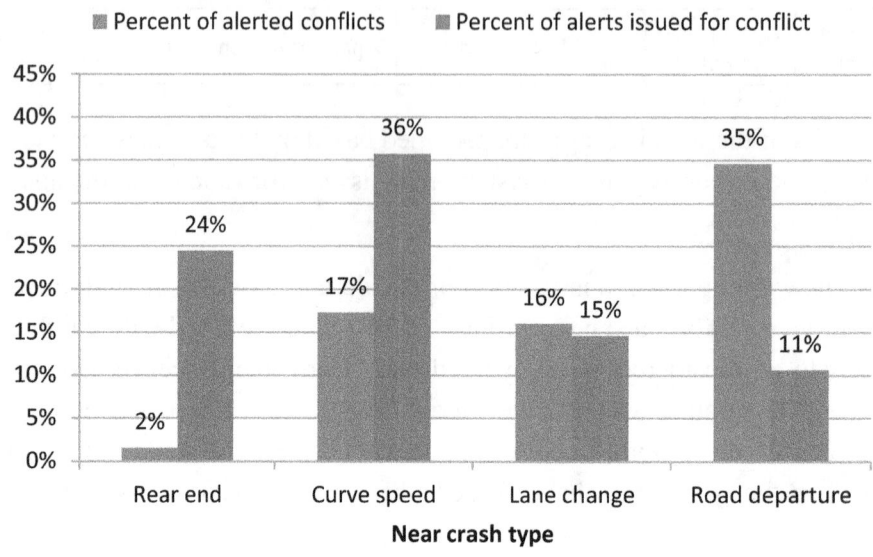

Figure 57. Mapping of valid alerts to driving conflicts

Figure 58 displays the results from a similar analysis that was conducted for near-crash events. The integrated safety system issued a valid alert in 467 (26 %) valid near crashes. Similar to the analysis of conflicts, rear-end near crashes had the lowest percent of alerted incidences and curve-speed warnings had the highest proportion of alerts issued for near crashes. Compared to the conflict results discussed above, percents are much higher for the number of alerted near crashes, and percent of alerts is much lower for near crashes. Since near crashes are characterized by higher intensity than driving conflicts, they are more likely to trigger an alert. Moreover, near crashes are more rare events than driving conflicts and therefore make up a smaller proportion of all alerts issued.

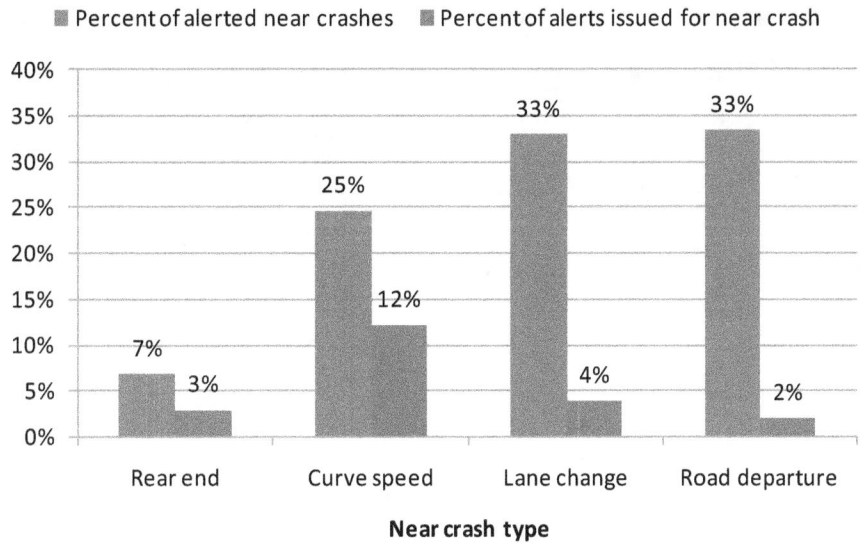

Figure 58. Mapping of valid near crashes to alerts

4.2.2 Driver Response

In addition to the assessment of sensor performance, data analysis was conducted to objectively infer whether or not system alerts impacted driver performance. In alert episodes with valid hazards, driver response was compared between the baseline period where drivers did not receive alerts and the treatment period. Driver response was expressed in terms of response type, brake reaction time and peak deceleration level to longitudinal alerts, and peak lateral acceleration to lateral alerts.

Figure 59 illustrates the action taken by drivers in response to alerts issued for valid threats based on video analysis. These data represent driver responses within five seconds after the alert and are based both on video analysis and numerical data. The magnitude of response metrics for each alert type is discussed below. For most alert types, drivers responded slightly more often when alerts were issued. Overall, drivers responded to 18 percent of valid alerts within five seconds of the alert onset.

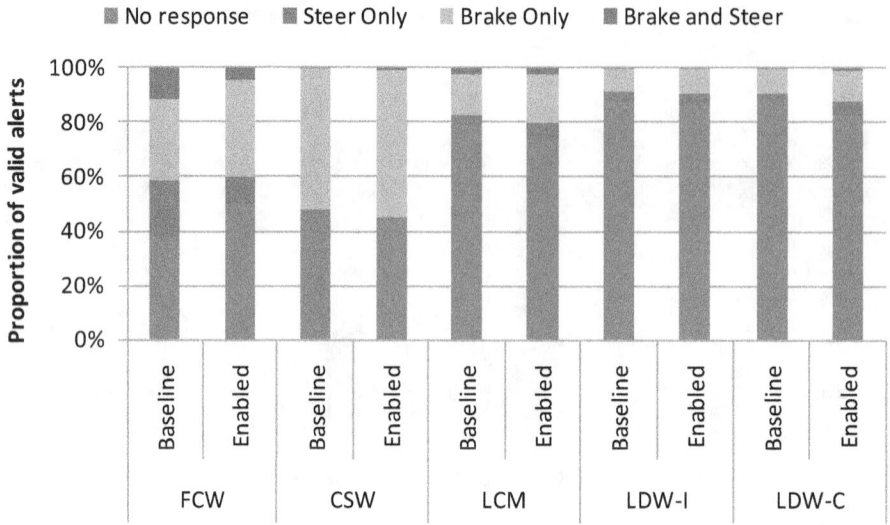

Figure 59. Breakdown of driver action in response to valid alerts

, Drivers generally responded to a longitudinal alert (FCW and CSW) by pressing the brake pedal. Brake pedal behavior was characterized in terms of reaction time and maximum deceleration. Reaction time is defined as the time interval between the onset of auditory alert and the moment the driver first activates the brake pedal. Reaction time quantifies how quickly drivers respond to alerts, while maximum deceleration portrays the intensity of the reaction.

Drivers showed a statistically significant change in response to forward-collision warnings both with respect to the speed and intensity of the response. During the baseline period, drivers activated the brake pedal an average of 1.03 seconds after the system triggered an alert (not audible to the driver). After the audible alerts were enabled, drivers activated the brake pedal an average of 0.58 seconds after the alert. Additionally, drivers decelerated more assertively when they received audible alerts, with an average maximum deceleration of 2.02 m/s^2 in treatment compared to 1.50 m/s^2 in baseline. As illustrated in Figure 60 and Figure 61, drivers respond to forward threats more quickly and more assertively when they received audible alerts from the integrated system.

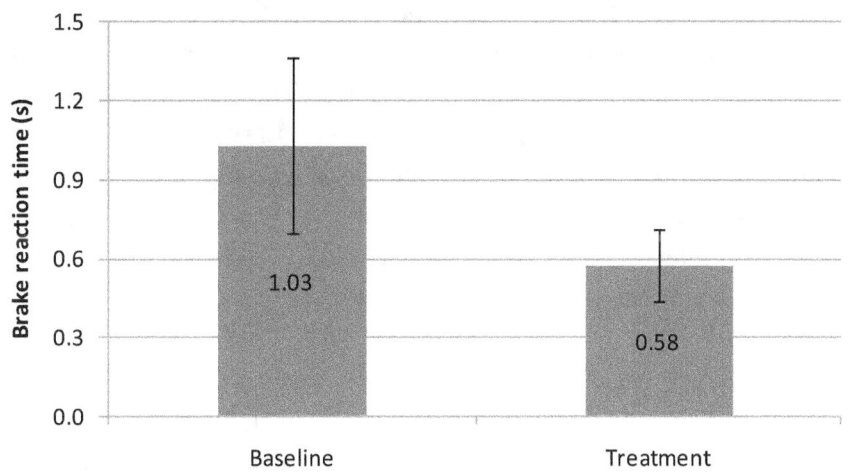

Figure 60. Average brake reaction time to FCW alerts in baseline and treatment

The analysis of driver response to curve-speed warnings shows a significant increase in deceleration when entering a curve when drivers receive an audible warning. Figure 61 illustrates that drivers had an average maximum deceleration of 1.43 m/s^2 in baseline as compared to 1.58 m/s^2 in treatment. Drivers also showed a slight reduction in response time to CSW alerts in the treatment period (0.84 seconds in baseline compared to 0.75 seconds during treatment), but this reduction was not statistically significant. With the integrated system enabled, drivers are more likely to reduce their speed when entering curves.

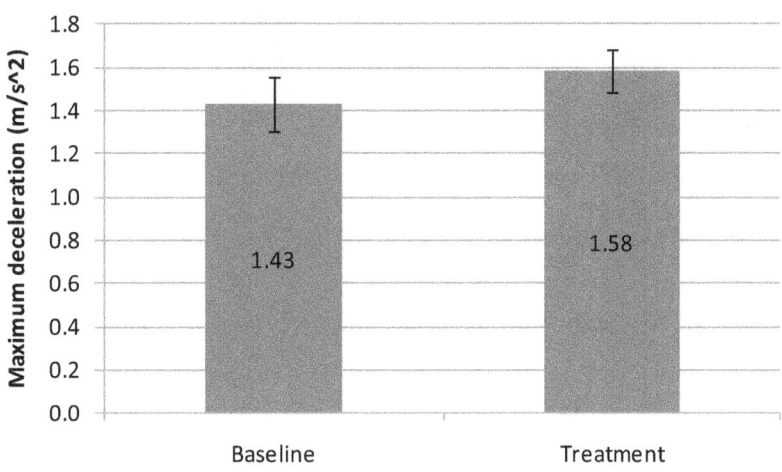

Figure 61. Average peak deceleration response CSW alerts in baseline and treatment

Drivers are most likely to respond to a lane-change warning or a drift warning with a steering correction. The driver response to these alerts was characterized by the maximum lateral speed and maximum lateral acceleration within five seconds after an alert was issued. These metrics define the timing and intensity of the driver's steering response.

The maximum lateral speed in response to a lateral warning gives insight into the timing of driver response to warnings. Figure 62 shows the average peak lateral speed for side alerts with valid threats in the baseline and treatment test conditions. The data show a statistically significant decrease in lateral speed after drivers receive an LDW-C alert from 1.57 m/s during baseline to 1.44 m/s after drivers received a seat vibration warning. If a driver begins a corrective movement earlier, there is less lateral distance to recover to resume lane position and the vehicle incurs lower lateral speeds. For cautionary drift warnings, drivers responded sooner to lane excursions when they felt the seat vibration alert.

Figure 63 shows the average peak lateral acceleration response for side alerts with valid threats. The data show a statistically significant increase in the steering intensity after drivers receive an LDW-C alert from 0.71 m/s^2 during baseline to 0.80 m/s^2 after drivers received a seat vibration warning. When the integrated system is enabled, drivers make more assertive steering responses to resume their lane position.

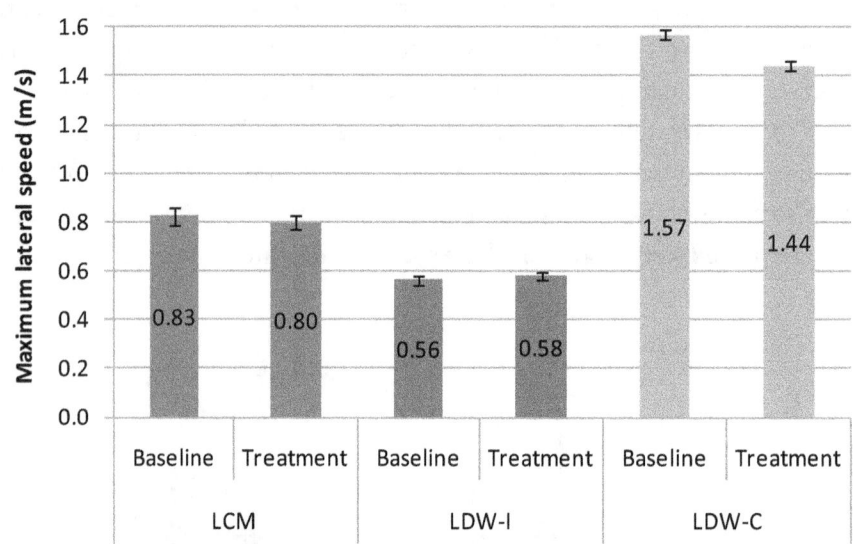

Figure 62. Average maximum lateral speed after lateral alerts

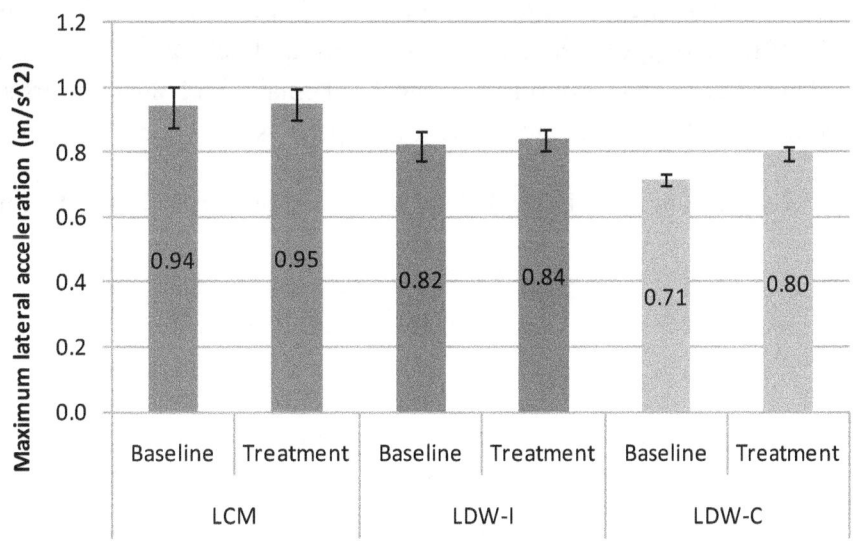

Figure 63. Average peak lateral acceleration after lateral alerts

4.2.3 Comparison of Alert Rates between Baseline and Treatment

The rate of exposure to alerts is representative of how drivers' behavior changes due to the integrated system. If drivers put themselves into fewer hazardous driving scenarios because of an increased alertness of their surroundings, they will receive fewer alerts. Table 28 shows the results of the paired t-test comparing the alert rates of the five subsystems and overall alert rates between treatment periods. The results show a significant reduction in both types of drift warnings for all drivers in both treatment periods. These results indicate that drivers are maintaining better lane positioning with the system enabled, and that these behavior changes are sustained after the initial behavior adaptation.

Table 28. Average number of alerts per 100 miles driven by treatment period

	Overall	FCW	CSW	LCM	LDW-I	LDW-C
Baseline vs. T_{all}:						
B	14.04	0.43	0.42	0.66	1.31	11.20
T_{all}	8.26	0.41	0.37	0.65	1.06	5.77
p	**0.00**	0.30	0.09	0.42	**0.00**	**0.00**
n	108	108	108	108	108	108
Baseline vs. T_2:						
B	14.04	0.43	0.42	0.66	1.31	11.21
T_2	8.16	0.40	0.37	0.65	1.10	5.64
p	**0.00**	0.27	0.12	0.42	**0.01**	**0.00**
n	108	108	108	108	108	108

Across all drivers, there was a 19 percent reduction in LDW-I alerts. Table 29 breaks down the LDW-I alert rate by age group, gender, and treatment period. Each age and gender group shows a trend towards fewer LDW-I alerts during T_{all}, and T_2. Females and older drivers showed a significant reduction in LDW-I alerts in both treatment periods.

Table 29. Average number of LDW-I alerts per 100 miles driven by treatment period

	Overall	Gender		Age (years)		
		Male	Female	20-30	40-50	60-70
Baseline vs. T_{all}:						
B	1.31	1.22	1.40	1.44	1.25	1.25
T_{all}	1.06	1.09	1.02	1.08	1.16	0.94
p	**0.00**	0.13	**0.00**	**0.00**	0.29	**0.01**
n	108	54	54	36	36	36
Baseline vs. T_2:						
B	1.31	1.22	1.40	1.44	1.25	1.25
T_2	1.10	1.07	1.13	1.26	1.07	0.97
p	**0.01**	0.12	**0.01**	0.06	0.14	**0.03**
n	108	54	54	36	36	36

Table 30 breaks down the LDW-C rate by age group, gender, and treatment period. Across drivers, there was a 49 percent reduction in LDW-C alert rate with the system enabled. The reduction was significant for both treatment periods for each demographic breakdown.

Table 30. Average number of LDW-C alerts per 100 miles driven by treatment period

	Overall	Gender		Age (years)		
		Male	Female	20-30	40-50	60-70
Baseline vs. T_{all}:						
B	14.04	13.15	9.27	10.62	11.45	11.57
T_{all}	8.26	6.49	5.05	5.40	5.15	6.77
p	**0.00**	**0.00**	**0.00**	**0.00**	**0.00**	**0.00**
n	108	54	54	36	36	36
Baseline vs. T_2:						
B	11.21	13.15	9.27	10.62	11.45	11.57
T_2	5.64	6.51	4.76	5.31	5.08	6.52
p	**0.00**	**0.00**	**0.00**	**0.00**	**0.00**	**0.00**
n	108	54	54	36	36	36

Cautionary lane-departure alert rates were further broken down by direction of departure. Sixty-six percent of all cautionary lane-departure warnings were issued for drifts to the left and 34 percent for drifts to the right. The reduction in LDW-C alert rate was statistically significant for departures to both the left and the right, as illustrated in Table 31. As with the overall reduction in the rate of LDW-C, the reductions were significant for both T_{all} and T_2 for all age and gender groups.

Table 31. Average number of LDW-C alerts per 100 miles driven by departure direction

	Left	Right
Baseline vs. T_{all}:		
B	7.23	4.29
T_{all}	3.89	1.90
p	**0.00**	**0.00**
n	108	108
Baseline vs. T_2:		
B	7.23	4.29
T_2	3.87	1.86
p	**0.00**	**0.00**
n	108	108

4.3 Driver-Vehicle Interface

Analysis of the driver-vehicle interface focused on the system display, auditory warnings, and system controls. The readability of visual information and the auditory alerts were evaluated through subjective feedback from field test participants gathered from the post-drive questionnaire and focus groups.

Figure 64 shows the results of three survey items related to the usefulness of the physical vehicle-driver interface. As mentioned in Section 1, the integrated system had a center display that showed a visual message when drivers received alerts, as well as two control buttons. One button adjusted the volume of the auditory alerts and the other allowed drivers to temporarily mute the auditory alerts. Of the physical elements of the driver interface, the center display received the highest proportion of positive scores. Only 13 percent of drivers rated the center display negatively. In the focus groups, some drivers commented that the center display would be more helpful if the message remained on the screen longer, and seven drivers said that their least favorite thing about the system was that the display was difficult to read. About half of the drivers said that they always looked to the center display when they received a warning to confirm the cause of the warning.

Responses were generally positive for the volume control and the mute buttons. About three quarters of the drivers responded positively to the usefulness of the volume control, and all but five drivers used the volume button. While 67 of the drivers reported that the mute button was useful, only 38 of the drivers took advantage of the mute button feature. Only nine drivers used the mute button regularly (more than five times).

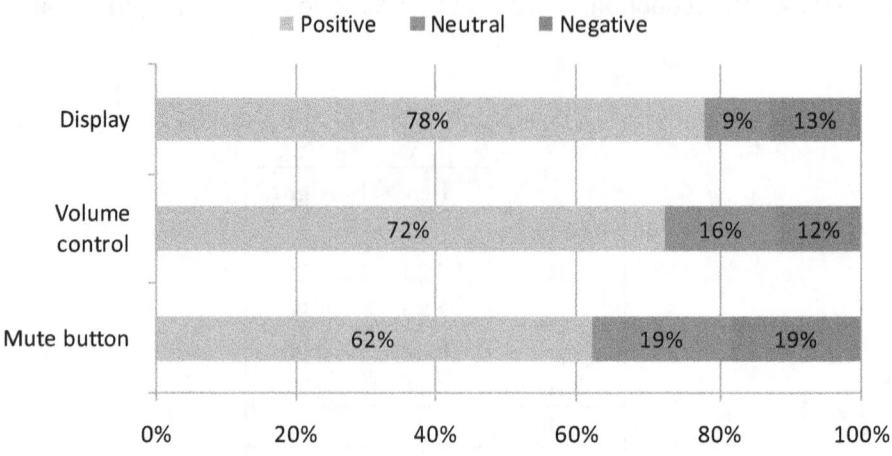

Figure 64. Drivers' reported usefulness of the physical driver-vehicle interface of the system

It is important that the alert modalities effectively get the attention of the drivers. Figure 65 shows drivers response to four questionnaire items that ask how attention-getting drivers found the different alert modalities. Drivers found the seat vibration to be the most attention getting. Responses on the brake pulse warning were mixed; some drivers commented that they did not even notice the brake pulse when it occurred, and others found it to be jarring and startling.

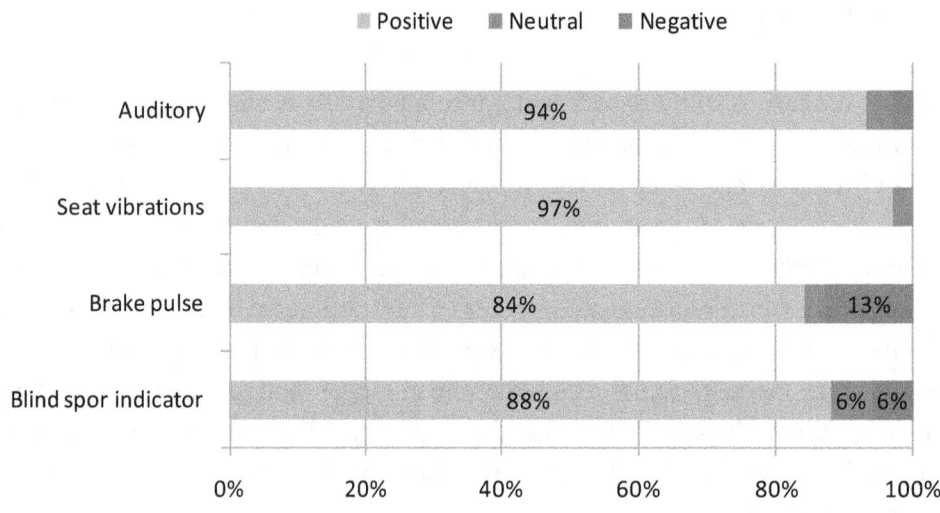

Figure 65. Drivers' opinions about the attention-getting capability of different elements of the driver-vehicle interface

4.4 System Robustness

System robustness was assessed by evaluating the availability of the LDW function of the system to issue a lane departure alert. System availability is a measure of the system's ability to recognize and track lane markers. It is important to note that the LDW function is disabled when the vehicle uses the turn signal or engages the brake pedal. The LDW function is considered "available" when it is able to recognize and track both lane markers. This enables the function to issue crash alerts for lateral drifting. Figure 66 illustrates the LDW availability under all driving conditions for left, right, and both sides of the travel lane for three different speed ranges. The lowest speed range (vehicle speeds between 25 and 35 mph) represents driving on local and rural roads. The middle speed range (between 35 and 55 mph) corresponds to driving on arterial roads. The upper speed range (over 55 mph) represents limited access highway driving. The LDW function was available when tracking markers on both sides of the lane for 88 percent of the mileage driven at speeds above 55 mph. This rate is higher than the other two speed ranges because highway lanes are generally better marked and maintained. The LDW availability function drops to 26 percent of the mileage traveled at lower speeds due to lower quality lane marking conditions on local and rural roads.

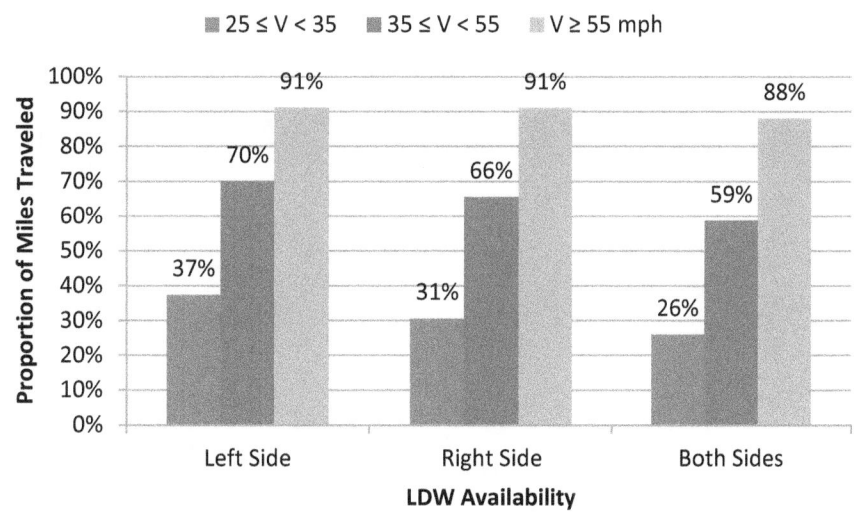

Figure 66. Availability of LDW function by travel speed

Table 32 summarizes the availability of the LDW function by travel speed under lighting and weather driving conditions. When tracking markers on both sides of the lane, the availability dropped by 22 percent at the lower speed bin, 13 percent at the middle speed bin, and six percent at the upper speed bin from daytime to nighttime conditions. From clear to adverse weather conditions, this performance decreased by 24 percent at the lower speed bin, 53 percent in the middle speed bin, and 78 percent in the upper speed bin.

Table 32. Availability of LDW function by travel speed and driving conditions

Travel Speed (mph)	Left Side				Right Side				Both Sides			
	Lighting		Weather		Lighting		Weather		Lighting		Weather	
	Day	Night	Clear	Adverse	Day	Night	Clear	Adverse	Day	Night	Clear	Adverse
$25 \leq V < 35$	26%	11%	36%	1%	22%	9%	29%	1%	18%	8%	25%	1%
$35 \leq V < 55$	50%	20%	67%	3%	47%	19%	62%	3%	42%	17%	56%	3%
$V \geq 55$	67%	24%	86%	5%	67%	24%	86%	5%	65%	23%	83%	5%

5. Conclusions

This section presents the key findings and discussion of the independent analysis of the IVBSS light vehicle field operational test including estimated crash reductions, reductions in driving conflicts and near crashes, positive changes in driver behavior, drivers' perceptions of the system, and sensor accuracy.

> **HIGHLIGHTS**
> - With 10 percent overall effectiveness, the integrated system could help prevent up to 788,000 light vehicle crashes annually
> - Drivers like driving with the system and felt that it would increase their driving safety
> - Drivers responded to threats more quickly and more assertively when they were issued system warnings

Safety Benefits:
- If all light vehicles in the United States were equipped with the integrated safety system, it is estimated that between 162,000 and 788,000 police-reported crashes could be prevented annually.
- The integrated system showed 40 percent effectiveness in reducing lane-change near crashes and 13 percent effectiveness in reducing road-departure near crashes.
- Drivers showed a significant increase in turn signal usage when making lane changes.
- Drivers showed a 21 percent decrease in the rate of lane busts with the system enabled, indicating improved lane keeping when driving with the system.
- For speeds over 55 mph, there was an overall decrease in conflict rate with the system enabled.
- The rate of lane-change conflicts decreased overall with the system enabled, especially conflicts to the left side; for conflicts to the right, the duration of those conflicts decreased.
- Road-departure conflicts on curved roads decreased with the system enabled and the duration of road-departure conflicts decreased on straight roads.
- No change in curve-speed conflict frequency was observed, but there was an increase in lateral acceleration during curve-speed conflicts.
- Fourteen of the 31 drivers who attended focus groups said that the integrated system helped prevent them from getting into a crash or near crash.
- Drivers did not show an increase in either secondary tasks or eyes-off-forward scene behavior with the system enabled, suggesting that the system did not have unintended consequences with respect to driver attention.
- All drivers showed a reduction in LCM and RD near crashes with the system enabled; the rate of LCM near crashes decreased more for men, and the rate of road departure conflicts decreased more for women.
- Younger drivers showed a 19 percent reduction in all near crashes with the system enabled.

Driver Acceptance:
- Eighty-two percent or drivers felt that the system would increase their driving safety.

- One third of drivers said that the integrated system issued nuisance warnings too frequently. Younger drivers were less tolerant of the nuisance warnings than middle-aged and older drivers; they were more likely to report that they received too many nuisance warnings, and more likely to find the nuisance warnings annoying.
- Seven drivers reported behavior adaptations that could potentially compromise their safety.
- Drivers' favorite feature of the integrated system was the blind spot monitors.
- Older drivers reported finding the system to be most useful and younger drivers found the system to be the least useful.
- Drivers found the lateral warning systems to be more useful and more desirable than the longitudinal warnings.
- Drivers reported exposure to false warnings was consistent with their actual exposure.

System Capabilities:
- Overall, alerts had a very high degree of accuracy.
- Alerts issued for forward stationary targets were issued mostly for out-of-path targets, indicating a low degree of accuracy for this type of FCW warning.
- Drivers respond to forward threats more quickly and more assertively when they received FCW alerts.
- Drivers showed more deceleration when approaching curves with the system enabled
- When the system was enabled, drivers made more assertive steering responses to resume their lane position.
- Drivers maintained better lane positioning with the system enabled (reduction in LDW warnings).
- With the system enabled, drivers showed a 46 percent reduction in drifts to the left, the type of drift that can lead to a head on collision.
- The system met the performance specification targets for lane tracking availability in all speed bins.

The system seemed to be a helpful tool for improving driving performance, decreasing exposure to both conflict and near-crash driving scenarios, and increasing overall driving awareness. The drivers using the system maintained better lane positioning, increased their use of their turn signals, and made more appropriate responses in threat scenarios. While some drivers reported unintended consequences no associated negative impact was observed.

Drivers reported a generally positive impression of driving with the integrated system, and most reported that they would like to have the system in their personal vehicle. Drivers felt that the system increased their safety and helped make them more aware of their surroundings. While most drivers were aware of the presence of nuisance warnings, the majority of drivers did not find the nuisance warnings to be annoying. Drivers cited the lateral warning systems as their favorites over the longitudinal systems, the blind spot monitors and the drift warnings being their favorite features. As a general rule, older drivers found the system to be more useful than the younger drivers.

The system alerts had a very high degree of accuracy, only issuing alerts when valid threats were present. The low response rate to valid alerts within five seconds of alerts (18 percent) suggests that many of the valid alerts may have been issued conservatively. Conservative warnings that do not require driver response could be perceived as nuisance warnings by the driver, but as mentioned above, most drivers did not feel they received too many nuisance warnings. One area of system performance with low accuracy was the accurate detection of forward stopped objects.

6. References

Ference, J.J., Szabo, S., and Najm, W.G. 2006. Performance Evaluation of Integrated Vehicle-Based Safety Systems. Proceedings of the Performance Metrics for Intelligent Systems (PerMIS) Workshop. Gaithersburg, MD: National Institute of Standards and Technology.

Lam, A.H., Bailin, A. Najm, W.G. and Nodine, E.E. 2009. Identification of Driving Conflicts in Light Vehicle Field Operational Test Data. Project Memorandum. HS22A1. Cambridge, MA: Volpe National Transportation Systems Center.

Najm, W.G. and J.D. Smith. 2007. Development of Crash Imminent Test Scenarios for Integrated Vehicle-Based Safety Systems (IVBSS). DOT HS 810 757. Washington, DC: National Highway Traffic Safety Administration.

Najm, W.G., Stearns, M.D., Howarth, H., Koopmann, J., and Hitz, J. 2006. Evaluation of an Automotive Rear-End Collision Avoidance System. DOT HS 810 569. Washington, D.C: National Highway Traffic Safety Administration.

Neale, V.L., Dingus, T.A., Klauer, S.G. Sudweeks, J., Goodman, M. 2005. An Overview of the 100- Car Naturalistic Study and Findings. TRB report number 05-0400, http://www-nrd.nhtsa.dot.gov/pdf/nrd-01/esv/esv19/05-0400-W.pdf.

Sayer, J., LeBlanc, D., Bogard, S., Nodine, E., and Najm, W.G. 2009. Integrated Vehicle-Based Safety Systems Third Annual Report. DOT HS 811 221. Washington, DC: National Highway Traffic Safety Administration.

Appendix A: Post-Drive Survey

IVBSS LV FOT Questionnaire and Evaluation

Please answer the following questions about the Integrated Vehicle Based Safety System (IVBSS). If you like, you may include comments alongside the questions to clarify your responses.

Example:
A.) Strawberry ice cream is better than chocolate.

1	2	3	4	5	6	7
Strongly Disagree						Strongly Agree

If you prefer chocolate ice cream over strawberry, you would circle the "1," "2" or "3" according to how strongly you like chocolate ice cream, and therefore disagree with the statement.

However, if you prefer strawberry ice cream, you would circle "5," "6" or "7" according to how strongly you like strawberry ice cream, and therefore agree with the statement.

If a question does not apply:

Write "NA," for "not applicable," next to any question which does not apply to your driving experience with the system. For example, you might not experience every type of warning the questionnaire addresses.

General Impression of the Integrated System

1. What did you like most about the integrated system?

2. What did you like least about the integrated system?

3. Is there anything about the integrated system that you would change?

4. How helpful were the integrated system's warnings?

| 1 | 2 | 3 | 4 | 5 | 6 | 7 |

Not all   Very
Helpful   Helpful

5. In which situations were the warnings from the integrated system helpful?

6. Overall, I think that the integrated system is going to increase my driving safety.

| 1 | 2 | 3 | 4 | 5 | 6 | 7 |

Strongly   Strongly
Disagree   Agree

7. Driving with the integrated system made me more aware of traffic around me and the position of my car in my lane.

| 1 | 2 | 3 | 4 | 5 | 6 | 7 |

Strongly   Strongly
Disagree   Agree

8. The integrated system made driving easier.

| 1 | 2 | 3 | 4 | 5 | 6 | 7 |

Strongly   Strongly
Disagree   Agree

9. Overall, I felt that the integrated system was predictable and consistent.

| 1 | 2 | 3 | 4 | 5 | 6 | 7 |

Strongly Disagree   Strongly Agree

10. I was not distracted by the warnings.

| 1 | 2 | 3 | 4 | 5 | 6 | 7 |

Strongly Disagree   Strongly Agree

11. Overall, how satisfied were you with the integrated system?

| 1 | 2 | 3 | 4 | 5 | 6 | 7 |

Very Dissatisfied   Very Satisfied

12. Did you rely on the integrated system? Yes____ No____

    a. If yes, please explain?

13. As a result of driving with the integrated system did you notice any changes in your driving behavior? Yes____ No____

    a. If yes, please explain.

14. Overall, I received warnings . . .

| 1 | 2 | 3 | 4 | 5 | 6 | 7 |

Too Frequently   Too Infrequently

If you answered Question 14 with a 1, 2, or 3, answer Question 14a below. If your answer was a 5, 6, or 7, answer Question 14b. If your answer was a 4, skip to Question 15.

a. If you received warnings too frequently, which type (s) of warnings did you receive too frequently? (circle all that apply)

Left/Right Hazard   Left/Right Drift   Hazard Ahead   Sharp Curve

b. If you received warnings too infrequently, which type (s) of warnings did you receive too infrequently? (circle all that apply)

Left/Right Hazard   Left/Right Drift   Hazard Ahead   Sharp Curve

15. I always understood why the integrated system provided me with a warning.

1   2   3   4   5   6   7

Strongly   Strongly
Disagree   Agree

16. I always knew what to do when the integrated system provided a warning.

1   2   3   4   5   6   7

Strongly   Strongly
Disagree   Agree

17. The auditory warnings' tones got my attention.

1   2   3   4   5   6   7

Strongly   Strongly
Disagree   Agree

18. I always understood why the integrated system provided me with an auditory warning tone.

1   2   3   4   5   6   7

Strongly   Strongly
Disagree   Agree

19. The auditory warnings' tones were not annoying.

| 1 | 2 | 3 | 4 | 5 | 6 | 7 |

Strongly Disagree   Strongly Agree

20. The seat vibration warnings got my attention.

| 1 | 2 | 3 | 4 | 5 | 6 | 7 |

Strongly Disagree   Strongly Agree

21. I always understood why the integrated system provided me with a seat vibration.

| 1 | 2 | 3 | 4 | 5 | 6 | 7 |

Strongly Disagree   Strongly Agree

22. The seat vibration warnings were not annoying.

| 1 | 2 | 3 | 4 | 5 | 6 | 7 |

Strongly Disagree   Strongly Agree

23. The brake pulse warnings got my attention.

| 1 | 2 | 3 | 4 | 5 | 6 | 7 |

Strongly Disagree   Strongly Agree

24. I always understood why the integrated system provided me with a brake pulse warning.

| 1 | 2 | 3 | 4 | 5 | 6 | 7 |

Strongly Disagree   Strongly Agree

25. The brake pulse warning was not annoying.

1   2   3   4   5   6   7

Strongly   Strongly
Disagree   Agree

26. The yellow lights in the mirrors got my attention.

1   2   3   4   5   6   7

Strongly   Strongly
Disagree   Agree

27. I always understood why the integrated system provided me with a yellow light in the mirror.

1   2   3   4   5   6   7

Strongly   Strongly
Disagree   Agree

28. The yellow lights in the mirrors were not annoying.

1   2   3   4   5   6   7

Strongly   Strongly
Disagree   Agree

29. Did you receive more than one warning within a few seconds (approximately three seconds)? Please place a check mark next to your answer.

Yes ____   No ____

30. The integrated system gave me warnings when I did not need them (i.e., nuisance warnings)

1   2   3   4   5   6   7

Strongly   Strongly
Disagree   Agree

31. Overall, I received nuisance warnings . . .

1	2	3	4	5	6	7
Too Frequently						Never

32. The nuisance warnings were not annoying.

1	2	3	4	5	6	7
Strongly Disagree						Strongly Agree

33. The integrated system gave me left/right hazard warnings when I did not need them.

1	2	3	4	5	6	7
Strongly Disagree						Strongly Agree

34. The integrated system gave me left/right drift warnings when I did not need them.

1	2	3	4	5	6	7
Strongly Disagree						Strongly Agree

35. The integrated system gave me hazard ahead warnings when I did not need them.

1	2	3	4	5	6	7
Strongly Disagree						Strongly Agree

36. The integrated system gave me sharp curve warnings when I did not need them.

1	2	3	4	5	6	7
Strongly Disagree						Strongly Agree

Overall Acceptance of the Integrated System

37. Please indicate your overall acceptance rating of the integrated system *warnings*
For each choice you will find five possible answers. When a term is completely appropriate, please put a check (√) in the square next to that term. When a term is appropriate to a certain extent, please put a check to the left or right of the middle at the side of the term. When you have no specific opinion, please put a check in the middle.

The integrated system **warnings** were:

useful	☐☐☐☐☐	useless
pleasant	☐☐☐☐☐	unpleasant
bad	☐☐☐☐☐	good
nice	☐☐☐☐☐	annoying
effective	☐☐☐☐☐	superfluous
irritating	☐☐☐☐☐	likeable
assisting	☐☐☐☐☐	worthless
undesirable	☐☐☐☐☐	Desirable
raising alertness	☐☐☐☐☐	sleep-inducing

Displays and Controls

38. The integrated system display was useful.

| 1 | 2 | 3 | 4 | 5 | 6 | 7 |

Strongly Disagree … Strongly Agree

39. The mute button was useful.

| 1 | 2 | 3 | 4 | 5 | 6 | 7 |

Strongly Disagree … Strongly Agree

40. The volume adjustment control was useful.

| 1 | 2 | 3 | 4 | 5 | 6 | 7 |

Strongly Disagree … Strongly Agree

41. Would you like to have the integrated system in your personal vehicle?

| 1 | 2 | 3 | 4 | 5 |

Definitely Not — Probably Not — Might or Might not — Probably Would — Definitely Would

42. What is the maximum amount that you would pay for the integrated system? Circle one price range.

- $0
- $250-500
- $500-750
- $750-1000
- $1000-1500
- $1500-2000
- More than $2000

Hazard Ahead Warning Acceptance

The Hazard Ahead warning provided an auditory warning accompanied by a brake pulse whenever you were approaching the rear of the vehicle in front of you and there was potential for a collision. When you received this type of warning, the display read "Hazard Ahead".

43. Please indicate your overall acceptance rating of the Hazard Ahead warnings.

For each choice you will find five possible answers. When a term is completely appropriate, please put a check (√) in the square next to that term. When a term is appropriate to a certain extent, please put a check to the left or right of the middle at the side of the term. When you have no specific opinion, please put a check in the middle.

The hazard ahead **warnings** when I was approaching a vehicle ahead were:

Useful	☐☐☐☐☐	useless
Pleasant	☐☐☐☐☐	unpleasant
Bad	☐☐☐☐☐	good
Nice	☐☐☐☐☐	annoying
Effective	☐☐☐☐☐	superfluous
Irritating	☐☐☐☐☐	likeable
Assisting	☐☐☐☐☐	worthless
undesirable	☐☐☐☐☐	Desirable
raising alertness	☐☐☐☐☐	sleep-inducing

Sharp Curve Warning Acceptance

The Sharp Curve warning provided an auditory warning whenever you were approaching a curve at too great a speed. When you received this type of warning, the display read "Sharp Curve".

44. Please indicate your overall acceptance rating of the Sharp Curve warnings.

For each choice you will find five possible answers. When a term is completely appropriate, please put a check (√) in the square next to that term. When a term is appropriate to a certain extent, please put a check to the left or right of the middle at the side of the term. When you have no specific opinion, please put a check in the middle.

The sharp curve **warnings** when I approached a curve at too great a speed were:

Useful	☐☐☐☐☐	useless
pleasant	☐☐☐☐☐	unpleasant
Bad	☐☐☐☐☐	good
Nice	☐☐☐☐☐	annoying
effective	☐☐☐☐☐	superfluous
irritating	☐☐☐☐☐	likeable
assisting	☐☐☐☐☐	worthless
undesirable	☐☐☐☐☐	Desirable
raising alertness	☐☐☐☐☐	sleep-inducing

Left/Right Hazard Warning Acceptance

The Left/Right Hazard warning provided an auditory warning whenever your turn signal was on AND you were changing lanes or merging and there was the possibility of a collision with a vehicle in the lane to which you were moving. Or, The Left/Right Hazard warning provided an auditory warning whenever your turn signal was not on and you were drifting out of your lane and there was the possibility of a collision with another vehicle or a solid object (e.g., a guard rail). When you received this type of warning, the display read "Left Hazard" or "Right Hazard" depending on your direction of travel.

45. Please indicate your overall acceptance rating of the Left/Right Hazard warnings.

For each choice you will find five possible answers. When a term is completely appropriate, please put a check (√) in the square next to that term. When a term is appropriate to a certain extent, please put a check to the left or right of the middle at the side of the term. When you have no specific opinion, please put a check in the middle.

The left/right hazard **warnings** were:

Useful	☐☐☐☐☐	useless
Pleasant	☐☐☐☐☐	unpleasant
Bad	☐☐☐☐☐	good
Nice	☐☐☐☐☐	annoying
Effective	☐☐☐☐☐	superfluous
Irritating	☐☐☐☐☐	likeable
Assisting	☐☐☐☐☐	worthless
Undesirable	☐☐☐☐☐	Desirable
raising alertness	☐☐☐☐☐	sleep-inducing

Left/Right Drift Warning Acceptance

If you were drifting out of your lane and there was no danger of you striking a solid object, you received a seat vibration and the display read "Left Drift" or "Right Drift" depending on the direction in which you were drifting.

46. Please indicate your overall acceptance rating of the Left/Right Drift warnings.

For each choice you will find five possible answers. When a term is completely appropriate, please put a check (√) in the square next to that term. When a term is appropriate to a certain extent, please put a check to the left or right of the middle at the side of the term. When you have no specific opinion, please put a check in the middle.

The left/right drift **warnings** were:

Useful	☐☐☐☐☐	useless
Pleasant	☐☐☐☐☐	unpleasant
Bad	☐☐☐☐☐	good
Nice	☐☐☐☐☐	annoying
Effective	☐☐☐☐☐	superfluous
Irritating	☐☐☐☐☐	likeable
Assisting	☐☐☐☐☐	worthless
undesirable	☐☐☐☐☐	Desirable
raising alertness	☐☐☐☐☐	sleep-inducing

Yellow Lights in the Mirrors Acceptance

When a vehicle was approaching or was in the research vehicle's blind spots, a yellow light in the exterior mirrors was illuminated.

47. Please indicate your overall acceptance rating of the yellow lights in the mirrors.

For each choice you will find five possible answers. When a term is completely appropriate, please put a check (√) in the square next to that term. When a term is appropriate to a certain extent, please put a check to the left or right of the middle at the side of the term. When you have no specific opinion, please put a check in the middle.

The **yellow lights** in the mirrors were:

Useful	☐☐☐☐☐	useless
pleasant	☐☐☐☐☐	unpleasant
bad	☐☐☐☐☐	good
nice	☐☐☐☐☐	annoying
effective	☐☐☐☐☐	superfluous
irritating	☐☐☐☐☐	likeable
assisting	☐☐☐☐☐	worthless
undesirable	☐☐☐☐☐	Desirable
raising alertness	☐☐☐☐☐	sleep-inducing

48. What is the maximum amount that you would pay for a system that warns you for hazards ahead? Circle one price range.

$0 $100-200 $200-300

$400-500 $500-750 $750-1000

More than $1000

49. What is the maximum amount that you would pay for a system that warns you when you are approaching a sharp curve too fast? Circle one price range.

$0 $100-200 $200-300

$400-500 $500-750 $750-1000

More than $1000

50. What is the maximum amount that you would pay for a system that warns you for drifting out of you lane? Circle one price range.

$0 $100-200 $200-300

$400-500 $500-750 $750-1000

More than $1000

51. What is the maximum amount that you would pay for a system that lets you know if you are about to make an unsafe lane change? Circle one price range.

$0 $100-200 $200-300

$400-500 $500-750 $750-1000

More than $1000

52. What is the maximum amount that you would pay for a system that lets you know if someone is in your blind spot? Circle one price range.

$0 $100-200 $200-300

$400-500 $500-750 $750-1000

More than $1000

Appendix B: Data Processing and Data Mining

Mining of the numerical data and coding of the video data collected from the field test and stored in a very large database are essential to the conduct of the independent evaluation. Data mining algorithms are developed to identify and categorize driving conflicts that map to target pre-crash scenarios. A video coding scheme and a data logger are created to compliment the MDAT in order to quantify visual information from the video data.

Data Mining

Data mining algorithms were developed to determine the occurrence of driving conflicts and near crashes in the field test (Lam, et al., 2009). The execution of these algorithms created new variables and data structures that were added to the independent evaluation database. The computed variables were developed based on the combination of measured parameters recorded by the DAS, mathematical computations, and-or equations from previous experimental projects. New data structures, implemented in a Microsoft SQL database, were designed to efficiently store massive amounts of driving data.

Data Mining Framework

The data processing framework consists of the following four steps that transform the raw field test data into aggregated data of driving conflicts:

- *Smooth and parse data*: This step smoothes the raw data by filling in very short gaps of missing data and filtering noisy data. This step makes the data easier to work with and makes results less erratic. Numerical algorithms for identifying vehicle maneuvers and driving conflicts are then run on the smoothed data to produce these new variables:

    - Vehicle maneuvers:
        o Vehicle states
        o Vehicle driving states
        o Vehicle maneuvers
        o Vehicle events
        o Driver responses
        o Lane keeping
        o Longitudinal, lateral, and combined motions
    - Driving conflicts:
        o Closing-in
        o Road and lanes departures
        o Changing lanes or merging

- *Identify significant events*: This step identifies significant events in the conflict driving states. This is followed by numerical analysis of the data to identify false driving conflicts, and-or using the MDAT to verify the occurrence of the conflicts.

- *Code events*: This step codes the significant events in a discrete variable database, after being stored as a continuous stream of sampled data from the previous step. This discrete database consists of conflict, vehicle, and driver files.

- *Aggregate events*: This final step queries the discrete database, using SQL or statistical programs, to aggregate all conflict events. The aggregated driving conflict data are then used by analysts to answer the independent evaluation questions.

Figure 67 and Figure 68 present the flowcharts of algorithms to identify respectively longitudinal and lateral driving conflicts based on raw field test data. The circular blocks represent the input data coming from the DAS. These data are drawn from the radar, in-vehicle, and sensors database. The green blocks denote the algorithms that produce the new variables and their concomitant data summary tables to be added to the independent evaluation database. Finally, the orange blocks refer to the conflict identification algorithms that use the variables created in post processing to determine whether or not a driving conflict has occurred. Rear-end driving conflicts are determined from the 50 percentile near crash threshold defined in (Najm et al., 2006). The rear-end/LVS scenario was filtered to exclude those based on a LVS event less than three seconds. Moreover, two consecutive longitudinal conflicts were counted as one conflict if they were separated by less than or equal to 2 seconds and had the same lead vehicle event. Specific thresholds used to determine conflicts are located in Appendix G. Table 33 lists the purpose of each of the data mining variables in the block diagrams below. Each variable is created to define a specific aspect of the driving scenario, which is ultimately used to determine the presence of conflicts. Variables are organized by what aspect of the driving scenario they define. Variables defining the host truck motion are created by in-vehicle data and lane tracking data. Forward target variables are derived primarily from forward-looking radar data. The variable that defines the side target location, adjacent target position, uses the side radar. The road geometry variable is derived from GPS map data. Each conflict variable is calculated using a combination of variables created in the data mining.

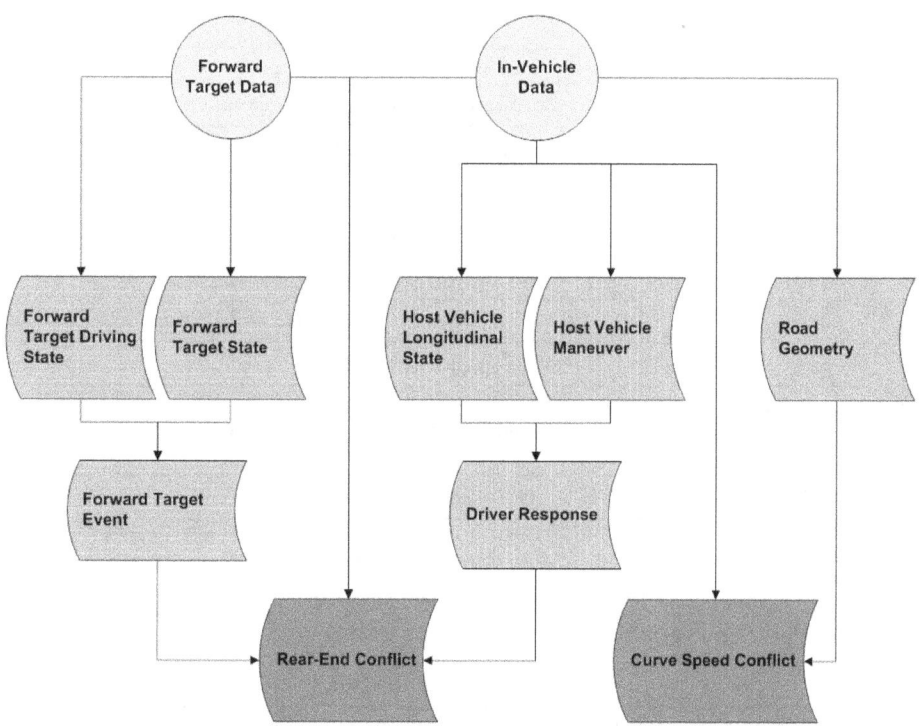

Figure 67. Block diagram of longitudinal driving conflicts

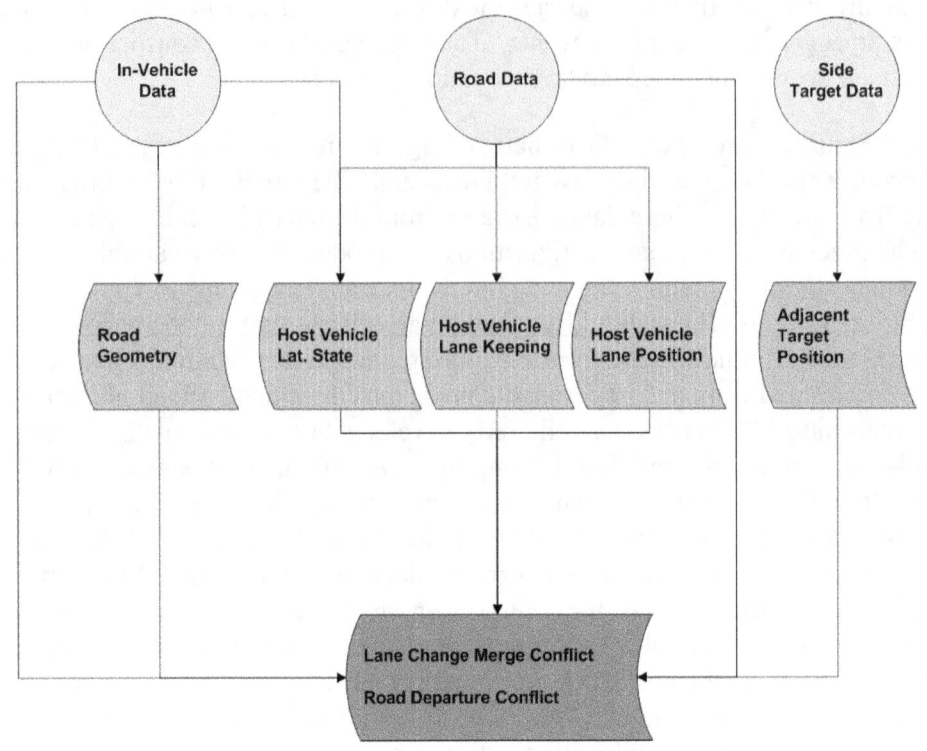

Figure 68. Block diagram of lateral driving conflicts

Table 33. Data mining variables

	Variable	Purpose
Host Truck	Host Vehicle Longitudinal State	Is truck accelerating, decelerating, or constant speed
	Host Vehicle Lane Keeping	Indentifies lane boundary violations
	Host Vehicle Lateral State	Lateral motion of truck
	Driver Response	Driver input to the truck
	Host Vehicle Maneuver	What is truck is doing (lane change, turning, going straight, etc)
Forward Target	In Path Target Count	Determines when the radar detects a new in-path target
	Forward Driving State	Relative speed of the lead vehicle
	Lead Vehicle State	Is lead vehicle accelerating, decelerating,constant speed, or stopped
	Lead Vehicle Category	
	Lead Vehicle Event	Defines events in in which the host vehicle is closing in on the target
Side Target	Adjacent Target Position	Defines relative location of side target
Road	Road Geometry	Determines road type and curvature of the road
Conflicts	Rear End Conflict	Identifies the presence of Rear end conflicts
	Lane Change/Merge Conflict	Identifies the presence of lane change/merge conflicts
	Road Departure Conflict	Identifies the presence of road departure conflicts

Appendix C: Video Analysis

A sample of 16,915 videos was selected for the analysis. Each video is tied to a system alert with a duration of 15 seconds, 10 seconds before to 5 seconds after the onset of the alert. This time frame encompasses time leading up to the alert to assess the driving scenario and time after the alert to gauge the driver's reaction to the event. All FCW, CSW, LCM, and LDW-I alerts were analyzed for all drivers. Due to the very large number of LDW-C alerts issued during the field test (17,186) a sample of each driver's LDW-C alerts from within the baseline and treatment periods was taken. The SQL random function was used to select a random sample of alerts proportionally from each treatment period for each driver.

Table 34 lists the total number of alerts and the number of analyzed alert videos by alert type. A total of 16,915 alert videos were analyzed, totaling 74 percent of all alerts.

Table 34. Breakdown of analyzed alert videos

	Number of alerts	Number of alerts analyzed	Percent of alerts analyzed
FCW	851	851	100%
CSW	919	919	100%
LCM	1,336	1,336	100%
LDW-I	2,501	2,501	100%
LDW-C	17,186	11,308	66%
Total	22,793	16,915	74%

Video events were coded to collect information that could only be obtained through video analysis of the alert episodes. The specific fields that were coded varied by alert type, based on the type of information that would be necessary to describe the type of driving scenario present for a specific type of alert. The variables are defined so that they can be coded with minimum subjectivity to create consistency in coding across alerts and different reviewers. Numerical data recorded by the DAS supplement the coded visual information for the analysis of alerts. Table 35 lists the variables that were coded for each alert type. See Appendix E for the coding manual that defines and quantifies the values for each of these variables.

Table 35. Variables coded in video analysis by alert type

All Alerts	FCW	CSW	LDW-I /LCM	LDW-I/LDW-C
Distraction	Target Type	Traverse Curve	Target Type	Lane Excursion Scenario
Eyes off Forward Scene	Target Vehicle Body Type	Passed Road Split	Target Location	Lane Marker
Steering Response	Lead Vehicle Maneuver		Moving Target Vehicle Speed	Road Condition
Host Vehicle Maneuver	Lead Vehicle Position			Opposing Traffic
Host Vehicle Position	In Path of Host Vehicle			Time of Collision
Location	Lead Vehicle Maneuver Times			

Appendix D: Video Coding Manual

INTRODUCTION

This section delineates the variables and codes that were derived from visual observation of video episodes captured during alerts issued by the integrated safety system during the field operational tests. The duration of each alert episode is 15 seconds – 10 seconds before alert onset and 5 seconds after. The following list of variables was created to collect information that can only be obtained through video analysis of the alert episodes. The variables are defined so that they can be coded with minimum subjectivity to create consistency in the coding across alerts and different reviewers. Numerical data from the data acquisition system will supplement the coded visual information for the analysis of alerts.

VARIABLES AND CODES

The following fields are to be entered based on observation of the video data.

I. All Alerts

The following fields are to be recorded for all alert types.

I.1. *Video Available*:
1. Yes
2. No
3. Not clear

I.2. *Crash Imminent*:
1. No
2. Yes
3. Unsure

I.3. *Distraction*:
1. None
2. Checking blind spot or rear view mirrors
3. Looking to the side/outside car
4. Grooming: High involvement
5. Grooming: Low involvement
6. Eating: Highly Involved
7. Eating: Low involvement
8. Drinking: Highly involved
9. Drinking: Low involvement
10. Adjusting controls
11. Adjusting/using aftermarket device
12. Dialing phone
13. Text messaging

14. Talking/listening to phone
15. Reading Cell Phone
16. Talking/listening to Bluetooth headset
17. Searching interior
18. Reaching for object in vehicle
19. Singing/whistling
20. Talking to/looking at passengers
21. Yawning
22. Eyes closed greater than 1 second
23. Smoking/lighting cigarette
24. Reading
25. Other
26. Unknown

I.4. *Eyes-Off-Forward-Scene*:
1. No (On road)
2. Yes (Off road)
3. Unsure

I.5. *Steering Response*:
1. None
2. Steering before alert
3. Steering after alert
4. Unsure

I.6. *Host Vehicle Maneuver*:
1. Going straight
2. Changing lanes
3. Turning
4. Merging
5. Negotiating curve
6. Other
7. Unsure

I.7. *Host Vehicle Position*:
1. Straight road
2. In curve
3. Curve entry
4. Curve exit
5. Other
6. Unsure

I.8. *Location*:
1. Normal Road
2. Ramp
3. Intersection

4. Normal road AND construction zone
5. Ramp AND construction zone
6. Intersection AND construction zone
7. Unsure

II. FCW Alerts

The following fields are to be recorded for FCW alerts only.

II.1. *Target Type*:
1. No target
2. Moving vehicle
3. Stationary vehicle
4. Roadside sign/object
5. Bridge/overhead sign
6. Guardrail/Jersey barrier
7. Embankment (earth or snow)
8. Pole
9. Other
10. Unknown

II.2. *Target Vehicle Body Type*:
1. No lead vehicle
2. Bicycle
3. Motorcycle
4. Compact/sedan/hatchback
5. SUV
6. Van or minivan
7. Light pick-up truck
8. Large truck
9. Bus
10. Other
11. Unsure

II.3. *Lead Vehicle Maneuver*:
1. No lead vehicle
2. Going straight
3. Cut in
4. Cut out
5. Cut in and out
6. Turning off
7. Turning across
8. Cut Across
9. Other
10. Unsure

II.4. *Lead Vehicle Position*:
1. No lead vehicle
2. Straight road
3. In curve
4. Curve entry
5. Curve exit
6. Unsure

II.5. *In Path of Host Vehicle*:
1. Yes
2. No
3. Unsure

II. 6. *Eyes on Forward Scene at Lead Vehicle Brake Onset Time*:
0. No
1. Yes
2. Unsure

II.7. *Eyes on Forward Scene at Lead Vehicle Cut-In Time*: Code if Lead Vehicle Maneuver is "Cut in," "cut in and out," "turning across" or "cut across".
0. No
1. Yes
2. Unsure

III. CSW Alerts

The following fields are to be recorded for CSW alerts only.

III.1. *Traverse Curve*:
1. No
2. Yes
3. Unsure

III.2. *Passed Road Split*:
1. No
2. Yes
3. Unsure

IV. LCM/LDW-I Alerts

The following fields are to be recorded for LCM and LDW-I alerts only.

IV.1. *Target Type*:
1. No target
2. Moving vehicle
3. Stationary vehicle

4. Roadside sign/object
   5. Bridge/overhead sign
   6. Guardrail/Jersey barrier
   7. Embankment (earth or snow)
   8. Pole
   9. Other
   10. Unknown

IV.2. *Target Location*:
   1. No target
   2. Adjacent
   3. Two or more lanes over
   4. Unsure
   5. Adjacent target in forward view

IV.3. *Moving Target Vehicle Relative Speed*:
   1. Faster
   2. Similar
   3. Slower
   4. No target vehicle
   5. Unknown

V. **LDW-C/LDW-I Alerts**

The following fields are to be recorded for LDW-C and LDW-I alerts only.

V.1. *Lane Excursion Scenario*:
   1. No excursion
   2. Intentional excursion/lane change
   3. Unintentional excursion
   4. Unsure

V.2. *Lane Marker*:
   1. Double
   2. Single-solid
   3. Single-dashed
   4. None/barely visible
   5. Unknown
   6. Other

V.3. *Road Condition*:
   1. Dry
   2. Wet
   3. Snow/slush
   4. Salt
   5. Unknown

6. Other

V.4 *Opposing Traffic* (left drift only)
1. No opposite direction lane
2. Clear opposite direction lane
3. Occupied opposite direction lane

V.5. ***Time-to-Collision*:** Number of data samples between the time the host vehicle first comes into contact with the lane boundary and the time that the vehicle overlaps with opposite direction vehicle in adjacent lane (left alerts only).

EXPLANATION AND CODING INSTRUCTIONS OF VARIABLES AND CODES

This section describes the variables and values presented above. It also provides instructions on how to determine the value for each variable.

I. **All Alerts**

I.1. *Video Available*:

This variable indicates whether or not a video is available for the alert.
1. Yes—All videos are available and clear
2. No—All 5 videos are missing. If some video channels are present and the others are missing, select "no" if the particular video/videos necessary to analyze the alert is missing (for example, forward video on an FCW alert).
3. Not clear—will be noted for episodes where the video is available and the scene is hard to "see" such as too dark at night.

I.2. *Crash Imminent*:

Watch the full length of the video, paying particular attention to the driver's reaction to the alert. Note whether or not the alert helped the driver avoid a collision and if the alert drew the driver's attention to the hazard. Use your judgment to determine if a crash was imminent at the time of the alert. In any instances of uncertainty, select "Unsure." For episodes coded as "Yes" or "Unsure," the results will be reviewed again by others in order to reach a consensus on the final code.

I.3. *Distraction Behavior*:

Pay particular attention to the driver's actions, face, and eyes in the 10 seconds leading up to the alert. Note any distractions that occur any time during this time period from the list below. Select up to 3 distraction behaviors.
1. None—no obvious distractions
2. Checking blind spot or rear-view mirrors—driver looks over shoulder or in mirrors
3. Looking to the side/outside cab—driver looks out windows
4. Grooming: Low involvement—driver scratches, runs fingers through hair, etc.

5. Grooming: Highly involved—driver applies makeup, using rearview mirror to look at himself, brushing hair, etc.
6. Eating: highly involved—driver unwraps food, eats a sandwich, etc.
7. Eating: Low involvement —driver eats candy, snacks, etc.
8. Drinking: Highly involved—driver opens drinks, tips bottle up to drink
9. Drinking: low involvement—driver sips through a straw, or sips etc.
10. Adjusting Controls—driver reaches towards center console to adjust in-vehicle controls
11. Adjusting/using aftermarket device—driver uses device such as navigation system or radar detector
12. Dialing phone—driver dials or presses buttons on his phone
13. Text messaging —driver presses buttons on his phone, but appears longer than dialing, or is not followed by talking
14. Talking/Listening to phone—phone visible, listening or talking
15. Reading cell phone—looking at cell phone but not dialing or talking
16. Talking/Listening to Bluetooth headset—earpiece is in, listening or talking
17. Searching interior—driver looks around interior of the vehicle, either front or back seat
18. Reaching for object in vehicle—driver retrieves object from somewhere in vehicle
19. Singing/whistling
20. Talking to/looking at passengers—driver engages in conversation with other occupants or looks at/is distracted by other occupants
21. Yawning
22. Eyes closed greater than 1 second—driver's eyes are visibly closed for a period of time longer than one second
23. Smoking or lighting cigarette—cigarette is visible, driver engages in any smoking-related behaviors, including opening window, ashing, smoking, opening cigarette box, etc.
24. Reading—reading material in view, eyes focused toward reading material
25. Other—any visible distraction that does not fit previous categories
26. Unknown—video not available

I.4. *Eyes-Off-Forward-Scene*:

Pay attention to the driver's gaze for the 5-second period before the alert. If the driver's eyes are focused anywhere other than the forward view for a period of at least 1.5 continuous seconds, the driver's eyes are considered to be "off the road." Select "unsure" if it is not possible to tell where the drivers gaze is directed.

I.5. *Steering Response*:

Using the forward view camera and the cabin camera, note whether the driver made any significant steering movements (larger than just minor corrections to remain on current track) just before or after the alert. If the steering correction was initiated at the same time as the alert onset, select "Steering before alert."

I.6. *Host Vehicle Maneuver*:

After watching the videos, make note of any intentional maneuver the driver performed immediately before the alert, or was performing during the time the alert was issued based on the driver's actions, and the front and side view videos. If more than one maneuver occurred, select the maneuver that you feel required the most driver attention. Also, more complicated maneuvers take precedence over less complicated ones. For example, if a driver is passing another vehicle while in a curve, select "Passing" rather than "on a curve."

1. Going straight: Driver travels on a straight road and remains in only one lane, without making any maneuvers.
2. Changing lanes: Driver executes a lane change on a multi-lane road. Directional signals may or may not be used.
3. Turning: Driver is turning or bearing off from one road to another.
4. Merging: Driver is merging into moving traffic on another road, or merging when a lane ends on their current road.
5. Negotiating curve: Vehicle is at any part of a sharp curve in the road, including highway exits or on-ramps and winding roads. This is the same as "going straight" but on a curved road.
6. Other: Other maneuver
7. Unsure: Not sure which maneuver the host vehicle is making

I.7. *Host Vehicle Position*:

Note the position of the host vehicle around the time of the alert.

1. Straight road: Vehicle is traveling on a straight road without intersecting roads.
2. In curve: Vehicle is navigating a curve.
3. Curve entry: Vehicle is approaching or just entering a curve.
4. Curve Exit: Vehicle is exiting a curve or has just completed the negotiation of a curve.
5. Unsure: Unsure of host vehicle position.

I.8. *Location*

Note the location of the host vehicle around the time of the alert.
1. Normal Road: vehicle is driving on a normal road (not a ramp) is not in an intersection and is not in a construction zone.
2. Ramp: Vehicle is navigating a highway on ramp or off ramp.
3. Intersection: Vehicle is passing through an intersection, or is approaching an intersection.
4. Normal road and Construction Zone: Vehicle is traveling through a construction zone where construction or multiple lane markings are visible.
5. Ramp and construction zone: Vehicle is traveling on a ramp that is also a construction zone.
6. Intersection and construction zone: Vehicle is traveling though an intersection where construction is also present.
7. Unsure: Unsure of vehicle location.

II. **FCW Alerts**

II.1. *Target Type*:

Watch the full length of the video paying particular attention to the forward scene and select the target that is most likely to have triggered the alert (most clearly in front of the vehicle). If the observed target is not on the list, select "Other." If it is not obvious what object caused the alert, select "Unknown." If no target is visible, select "None".

II.2. *Target Vehicle Body Type*:

Identify the type of moveable target that triggered the alert. If "Target type" is not "moving vehicle" select "no lead vehicle".

II.3. *Lead Vehicle Maneuver*:

If the "Target type" has been noted as "Moving Vehicle," note any maneuvers the lead vehicle is making at the time of the alert.
1. No lead vehicle: "Target type" is not "Moving Vehicle".
2. Going Straight: Lead vehicle is traveling in its current lane without making any maneuvers.
3. Cut in: Lead vehicle executes a lane change from an adjacent lane into the lane of travel of the host vehicle, or lead vehicle turns onto roadway in front of host vehicle. Lead vehicle may cut in from the other direction (of what is shown below):

4. Cut out: Lead vehicle executes a lane change to adjacent lane so that they are no longer in the same lane of travel of the host vehicle. Lead vehicle may cut out to the other direction (of what is shown below):

5. Cut in and out: Lead vehicle executes a cut out immediately after a cut in; i.e., moves from one adjacent lane to the adjacent lane on the other side of the vehicle. Lead vehicle may execute this to the other direction (of what is shown below):

   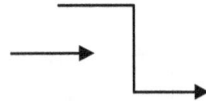

6. Turning off: Lead vehicle is preparing to turn onto another road (is slowing), or is turning onto another road. Use blinker to help determine if the lead vehicle intends to turn. Lead vehicle may turn into the other direction (of what is shown below):

7. Turning across: Target vehicle is turning onto a perpendicular road from opposite direction of travel, and passes across path of host vehicle.

8. Cut Across: Target vehicle is traveling across (perpendicular to) the host vehicle's lane of travel at an intersection. Lead vehicle may cut across the other direction (of what is shown below):

9. Unsure: Target

II.4. *Lead Vehicle Position*:

From the forward scene video, determine the characteristics of the road where the lead vehicle is at the time of the alert.
1. No lead vehicle: "Target type" is not coded as "moving vehicle"
2. Straight road: Lead vehicle is traveling on a straight road.
3. In curve: Lead vehicle is navigating a curve.
4. Curve Entry: Lead vehicle is just entering a curve.
5. Curve Exit: Lead vehicle is completing the negotiation of the curve.
6. Unsure

II.5. *In Path of Host Vehicle*:

This variable denotes whether the target is or was in the intended path of the equipped host vehicle (in the lane of travel of the host vehicle) around the alert time. If the vehicle is currently in path at the alert time, or if the vehicle cut in or out of the equipped vehicle path, code as "Yes."

II. 6. *Eyes on Forward Scene at Lead Vehicle Brake Onset Time*:

Note whether the driver's attention was on the forward scene at the time, or within 2 samples of the lead vehicle's brake onset. Leave blank if there is no brake onset time. Enter the appropriate number into the entry field.
0. No: Driver's attention is not on forward scene
1. Yes: Driver's attention is on forward scene
2. Unsure

II.7. *Eyes on Forward Scene at Lead Vehicle Cut-In Time*:

Note whether the driver's attention was on the forward scene at the time, or within 2 samples of the time the lead vehicle first begins to enter the host vehicle's lane. Code if Lead Vehicle Maneuver is "Cut in," "cut in and out," "turning across" or "cut across," otherwise leave blank . Enter the appropriate number into the entry field.
0. No: Driver's attention is not on forward scene
1. Yes: Driver's attention is on forward scene
2. Unsure

III. CSW Alerts

III.1. *Traverse Curve*:

From the forward scene, determine if the host vehicle traverses the curve (before, during, or after the alert) for which the warning was issued.
1. Yes: Host vehicle was traversing or entering a curve at the time of the CSW alert.
2. No: Host vehicle was not traversing a curve at the time of, or shortly after a CSW alert.
3. Unsure

III.2. *Passed Road Split*:

If the host vehicle did not traverse a curve, indicate whether or not the vehicle passed a road split where a curve was present (e.g., off ramp)
1. No: split in road
2. Yes: Passed split in road where a curve was present
3. Unsure

IV. LCM/LDW-I Alerts

IV.1. *Target Type*:

Watch the full length of the video paying particular attention to the side scene videos and select the target that most likely to have triggered the alert. If the observed target is not on the list, select "Other." If it is not obvious what object caused the alert, select "Unknown." If no target is visible, select "None."

IV.2. *Target Location*:

The position of side targets with respect to the equipped vehicle at the time of the alert. If no side target or target is unidentifiable, select "No Target." If the equipped vehicle or a POV is changing lanes at the time of the alert, select the lane the vehicle was in before the maneuver.
1. No target
2. Adjacent: Target is adjacent to host vehicle in either a lane of travel, road shoulder, or off the road.
3. Two or more lanes over: There is a full travel lane between the host vehicle and the target.
4. Unknown: Unable to determine the lateral offset of the target.
5. Adjacent target in forward view: Target is in adjacent lane, but is in front of the host vehicle. Rear bumper of vehicle must be visible in the forward camera.

V1.3. *Moving Target Vehicle Relative Speed*:

Note the speed of the lateral moving target relative to the host vehicle at the time of the alert. Use the side and front cameras to determine the relative speed over the length of the video.
1. The target vehicle is traveling faster than the host vehicle.

2. The target is traveling approximately the same speed as the host vehicle.
3. The host vehicle is passing the target vehicle.
4. No target vehicle
5. The relative speed is not determined, or inconsistent over the course of the video.

V. LDW-C/LDW-I Alerts

V.1. *Lane Excursion Scenario*:

By watching the forward and side view videos, indicate the lane keeping behavior of the host vehicle:
1. No Excursion: Vehicle did not leave lane or drift towards lane edge.
2. Intentional Excursion: Driver intentionally swerved out of or to the side of their lane to avoid another vehicle, pedestrian, bicycle, an object in the roadway or driver changes lanes intentionally without turn signal or driver cuts curve to make wide turn or maneuver.
3. Unintentional Excursion: Vehicle leaves lane or drifts towards lane edge, apparently unintentionally, or showing no signs of an intentional maneuver.
4. Unsure: Unclear whether vehicle departed lane or drifted in lane

V.2. *Lane Marker*:

Using the side and forward view cameras, determine the type of lane markings relevant to the side of the alert. Select unknown if the lane marking is undetermined because of poor video quality.

V.3. *Road Condition*:

Using the front view video, indicate the condition of the road surface when alert is issued.
1. Dry: No visible moisture or residue on road
2. Wet: Visible moisture or standing water
3. Snow or slush: Accumulating snow or slush on roadway
4. Salt: Visible salt residue on roadway, possibly obstructing lane lines
5. Unknown: Not able to determine.

V.4. *Opposing Traffic* (left drift only)

This variable makes note of whether a vehicle is approaching in an adjacent, opposite-direction travel lane. **This variable should only be coded for left drift alerts.** A lane is considered to have an adjacent opposite-direction travel lane only if it is the leftmost travel lane in that direction, and if the road has no barrier (curb, grassy area, Jersey barrier, etc) between the opposite direction lanes.

1. No opposite direction lane – travel lane is not the leftmost travel lane, or there is a divider (curb, grass, jersey barrier etc.) between the opposite direction roadways.
2. Clear opposite direction lane – lane directly to the left of the travel lane is an opposite direction lane, lane is unoccupied.

3. Occupied opposite direction lane— lane directly to the left of the travel lane is an opposite direction lane, lane has approaching vehicle at any time 10s before or 5s after the alert onset.

V.5. Time-to-Collision:

If "Opposing Traffic" is coded "occupied opposite direction lane," count the number of samples between the time the host vehicle first comes in contact with the lane boundary, and the time that the vehicles first meet, or the time their bumpers would come into contact if they were occupying the same lane. Enter the number of samples into the entry field. If the opposing lane vehicle does not meet the host vehicle before the end of the video, leave this field blank.

Appendix E: Overall Driving Analysis Supplemental Data

Table 36: Lane busts per 100 miles driven

	Overall	Gender		Age (years)			Road Type		Speed (mph)			
		Male	Female	20-30	40-50	60-70	Freeway	Non-Freeway	25-35	35-45	45-55	55+
Baseline vs. T_{all}:												
B	38.7	37.1	40.3	41.1	40.4	34.5	20.6	55.9	81.0	55.2	39.9	19.9
T_{all}	30.6	29.2	32.0	33.2	29.6	29.0	15.4	44.9	70.9	44.9	30.0	13.7
p	**0.00**	**0.00**	**0.00**	**0.00**	**0.00**	**0.01**	**0.00**	**0.00**	**0.00**	**0.00**	**0.00**	**0.00**
n	108	54	54	36	36	36	108	108	108	108	108	108
Baseline vs. T_2:												
B	38.7	37.1	40.3	41.1	40.4	34.5	20.6	55.9	81.0	55.2	39.9	19.9
T_2	31.1	30.2	32.0	34.1	29.5	29.7	16.0	45.5	71.6	46.0	30.7	14.4
p	**0.00**	**0.00**	**0.00**	**0.01**	**0.00**	**0.02**	**0.00**	**0.00**	**0.00**	**0.00**	**0.00**	**0.00**
n	108	54	54	36	36	36	108	108	108	108	108	108

Table 37: Lane-bust duration (sec)

	Overall	Gender		Age (years)			Road Type		Speed (mph)			
		Male	Female	20-30	40-50	60-70	Freeway	Non-Freeway	25-35	35-45	45-55	55+
Baseline vs. T_{all}:												
B	2.72	2.70	2.74	2.72	2.72	2.72	2.51	2.81	3.08	2.82	2.67	2.49
T_{all}	2.64	2.59	2.69	2.70	2.56	2.65	2.46	2.76	3.12	2.70	2.67	2.43
p	**0.03**	**0.02**	0.38	0.61	**0.01**	0.44	0.43	0.21	0.50	**0.04**	0.98	0.40
n	108	54	54	36	36	36	108	108	108	108	108	108
Baseline vs. T_2:												
B	2.72	2.70	2.74	2.72	2.72	2.72	2.51	2.81	3.08	2.82	2.67	2.49
T_2	2.66	2.64	2.68	2.75	2.56	2.68	2.46	2.80	3.17	2.70	2.67	2.47
p	0.17	0.32	0.35	0.66	**0.03**	0.63	0.54	0.97	0.24	**0.04**	0.99	0.80
n	108	54	54	36	36	36	108	108	108	108	108	108

Table 38: Average vehicle speed (m/s) three seconds prior to curve start

	Overall	Gender		Age (years)			Road Type		Min. Curve Radius (m)		
		Male	Female	20-30	40-50	60-70	Freeway	Non-Freeway	100-250	250-500	500-1000
Baseline vs. T_{all}:											
B	22.4	23.2	21.5	22.5	23.4	21.2	27.4	18.8	18.4	22.7	25.6
T_{all}	22.5	23.1	21.8	22.8	23.4	21.2	27.6	18.9	18.4	22.8	25.7
p	0.70	0.68	0.38	0.40	0.87	1.00	0.72	0.28	0.93	0.76	0.69
n	104	54	50	35	35	34	104	104	104	104	104
Baseline vs. T_2:											
B	22.4	23.2	21.5	22.5	23.4	21.2	27.4	18.8	18.4	22.7	25.6
T_2	22.7	23.2	22.0	23.0	23.5	21.5	28.1	19.0	18.6	22.9	25.8
p	0.25	0.93	0.16	0.27	0.89	0.57	0.20	0.18	0.45	0.45	0.54
n	104	54	50	35	35	34	104	104	104	104	104

Appendix F: Conflict Identification Thresholds

The conditions used to determine conflicts analyzed in this document are presented below by conflict type. More details about the data mining procedures used for the data in these analyses can be found in (Najm et al., 2007).

Rear-End Driving Conflicts

Three types of rear-end conflicts are included in this analysis: lead vehicle stopped, lead vehicle decelerating, and lead vehicle moving at slower constant speed. For a conflict to be present, the following criteria must be satisfied:

- Forward target is present (stopped, decelerating, or constant speed)
- Driver response present (braking or steering)
- Time-to-collision TTC and range rate within thresholds shown below per lead vehicle state and driver response.

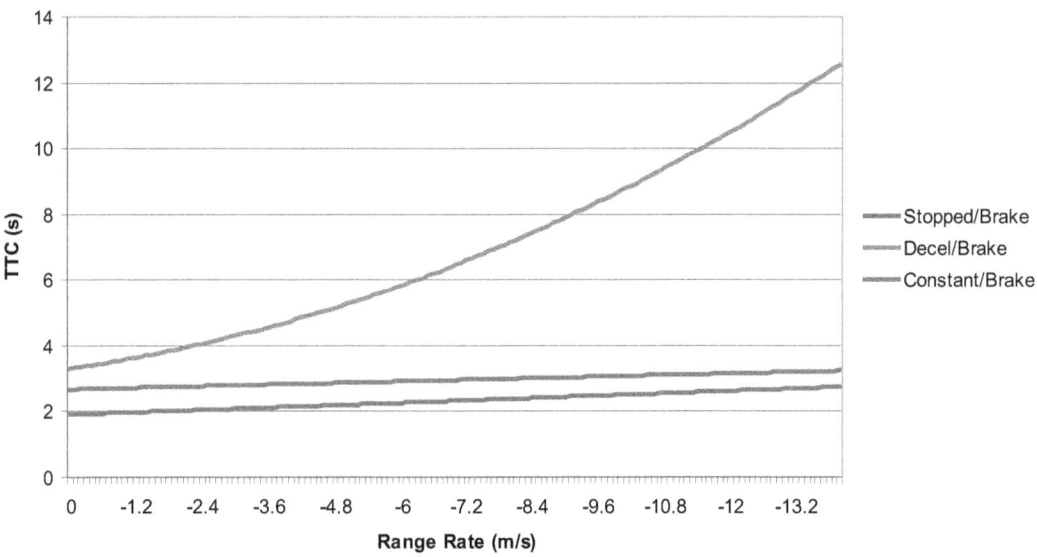

Curve-Speed Driving Conflict

Four conditions must be met to determine that a curve-speed conflict has occurred:
- Vehicle is approaching or negotiating a curve
- Speed is greater than the calculated maximum speed
- Maximum required deceleration is less than -0.11
- Lateral acceleration is greater than 0.4 g

Lane-Change Driving Conflict

The following criteria must be met to determine that a lane-change conflict has occurred:

- Target is present
- Lane boundary is dashed (not solid)

- Counter steering response present
- Lateral acceleration response (in direction back into lane) above 0.75 m/s^2 on a straight road (on curved road, lateral acceleration threshold varies with road geometry)

Road-Departure Driving Conflict

The following criteria must be met for a driving scenario to be considered a road-departure conflict:

- Vehicle crosses solid boundary
- Counter steering response present
- Lateral acceleration response (in direction back into lane) above 1.5 m/s^2 on a straight road (on curved road, lateral acceleration threshold varies with road geometry)

Appendix G: Driving Conflict Analysis Supplemental Data

Table 39. Overall number of conflicts per 100 miles driven

	Overall	Gender		Age (years)			Light		Weather		Speed (mph)	
		Male	Female	20-30	40-50	60-70	Night	Day	Clear	Adverse	25-55	55+
Baseline vs. T_{all}:												
B	10.13	9.89	10.36	11.55	9.18	9.65	8.19	11.36	10.40	9.37	8.39	4.14
T_{all}	10.36	9.88	10.85	11.00	9.53	10.57	8.23	11.82	10.60	8.96	8.34	3.71
p	0.52	0.98	0.42	0.32	0.42	0.29	0.93	0.33	0.60	0.77	0.89	**0.03**
n	108	54	54	36	36	36	90	108	108	55	106	99
Baseline vs. T_2:												
B	10.13	9.89	10.36	11.55	9.18	9.65	8.17	11.36	10.40	9.25	8.45	4.14
T_2	10.26	9.92	10.60	10.62	9.72	10.43	7.79	11.63	10.45	9.72	8.75	3.69
p	0.75	0.96	0.72	0.20	0.30	0.38	0.34	0.62	0.90	0.78	0.48	0.07
n	108	54	54	36	36	36	88	108	108	48	104	99

Table 40. Average number of rear-end conflicts per 100 miles driven

	Overall	Gender		Age (years)			Light		Weather		Speed (mph)	
		Male	Female	20-30	40-50	60-70	Night	Day	Clear	Adverse	25-55	55+
Baseline vs. T_{all}: Lead vehicle decelerating (LVD)												
B	3.53	3.26	3.80	3.78	3.13	3.69	3.15	4.08	3.63	3.64	3.00	-
T_{all}	3.76	3.30	4.22	3.95	3.34	3.99	2.82	4.49	3.84	4.19	3.03	-
P	0.21	0.85	0.17	0.49	0.38	0.49	0.13	0.08	0.26	0.44	0.87	-
n	107	54	53	35	36	36	62	105	107	33	99	-
Baseline vs. T_2: Lead vehicle decelerating (LVD)												
B	3.54	3.26	3.82	3.78	3.13	3.71	3.10	4.10	3.63	3.46	3.00	-
T_2	3.74	3.36	4.13	3.75	3.41	4.07	2.86	4.46	3.82	4.38	3.15	-
P	0.29	0.70	0.30	0.91	0.28	0.43	0.23	0.16	0.35	0.24	0.47	-
n	106	54	52	35	36	35	58	104	106	23	96	-

	Overall	Gender		Age (years)			Light		Weather		Speed (mph)	
		Male	Female	20-30	40-50	60-70	Night	Day	Clear	Adverse	25-55	55+
Baseline vs. T_{all}: Lead vehicle moving at slower constant speed (LVM)												
B	0.78	0.73	0.83	1.11	0.58	0.65	1.18	0.95	0.82	-	0.94	1.19
T_{all}	0.82	0.79	0.87	1.11	0.67	0.68	0.78	1.02	0.85	-	0.80	0.87
p	0.43	0.44	0.72	0.97	0.28	0.70	**0.03**	0.28	0.59	-	0.22	0.14
n	86	47	39	29	30	27	29	79	84	-	45	19
Baseline vs. T_2: Lead vehicle moving at slower constant speed (LVM)												
B	0.80	0.74	0.87	1.14	0.61	0.63	1.23	0.97	0.83	-	1.01	1.25
T_2	0.86	0.80	0.93	1.09	0.73	0.74	0.91	1.09	0.88	-	1.04	0.92
p	0.45	0.42	0.69	0.80	0.26	0.19	0.13	0.28	0.57	-	0.89	0.15
n	81	46	35	28	28	25	25	73	80	-	37	17

Table 41. Average response intensities to rear-end conflicts

Time to collision at brake onset (seconds):

		Overall	Gender		Age (years)			Speed (mph)	
			Male	Female	20-30	40-50	60-70	25-55	55+
LVD	*Baseline vs. T_{all}*:								
	B	10.70	10.94	10.44	9.98	11.27	10.81	8.73	-
	T_{all}	10.72	10.62	10.83	10.40	10.91	10.84	9.34	-
	p	0.95	0.48	0.52	0.55	0.54	0.97	0.30	-
	n	105	54	51	34	36	35	94	-
	Baseline vs. T_2:								
	B	10.70	10.94	10.44	9.98	11.27	10.81	8.43	-
	T_2	10.53	10.46	10.61	10.12	10.84	10.62	9.86	-
	p	0.65	0.31	0.77	0.83	0.50	0.78	0.09	-
	n	104	54	50	34	36	34	91	-
LVM	*Baseline vs. T_{all}*:								
	B	3.31	3.23	3.41	3.25	3.29	3.38	3.51	-
	T_{all}	3.29	3.28	3.30	3.37	3.33	3.17	3.59	-
	p	0.86	0.53	0.49	0.47	0.83	0.07	0.66	-
	n	84	46	38	28	29	27	38	-
	Baseline vs. T_2:								
	B	3.32	3.23	3.44	3.28	3.32	3.36	3.51	-
	T_2	3.39	3.28	3.53	3.39	3.47	3.29	3.57	-
	p	0.60	0.70	0.72	0.64	0.54	0.69	0.79	-
	n	77	44	33	26	26	25	31	-

Minimum time to collision during conflict resolution (seconds):

		Overall	Gender		Age (years)			Speed (mph)	
			Male	Female	20-30	40-50	60-70	25-55	55+
LVD	*Baseline vs. T_{all}*:								
	B	2.68	2.69	2.68	2.63	2.68	2.73	2.95	-
	T_{all}	2.68	2.68	2.69	2.64	2.68	2.73	2.92	-
	p	0.97	0.86	0.84	0.88	0.94	0.89	0.55	-
	n	107	54	53	35	36	36	97	-
	Baseline vs. T_2:								
	B	2.68	2.69	2.67	2.63	2.68	2.72	2.95	-
	T_2	2.70	2.70	2.71	2.64	2.71	2.76	2.98	-
	p	0.44	0.88	0.36	0.94	0.61	0.48	0.61	-
	n	106	54	52	35	36	35	94	-
LVM	*Baseline vs. T_{all}*:								
	B	2.52	2.49	2.55	2.51	2.51	2.53	2.56	-
	T_{all}	2.50	2.51	2.50	2.49	2.55	2.46	2.61	-
	p	0.65	0.59	0.20	0.69	0.29	0.11	0.35	-
	n	83	45	38	28	29	26	37	-
	Baseline vs. T_2:								
	B	2.51	2.49	2.54	2.52	2.50	2.52	2.57	-
	T_2	2.48	2.52	2.43	2.44	2.56	2.43	2.59	-
	p	0.38	0.48	**0.04**	0.25	0.25	0.19	0.76	-
	n	76	43	33	26	26	24	28	-

Peak deceleration level during conflict resolution (m/s^2):

		Overall	Gender		Age (years)			Speed (mph)	
			Male	Female	20-30	40-50	60-70	25-55	55+
LVD	*Baseline vs. T$_{all}$:*								
	B	-2.53	-2.57	-2.48	-2.54	-2.55	-2.49	-2.51	-
	T$_{all}$	-2.53	-2.55	-2.51	-2.55	-2.57	-2.47	-2.59	-
	p	0.90	0.76	0.65	0.88	0.78	0.78	0.35	-
	n	107	54	53	35	36	36	97	-
	Baseline vs. T$_2$:								
	B	-2.53	-2.57	-2.48	-2.54	-2.55	-2.49	-2.50	-
	T$_2$	-2.54	-2.55	-2.52	-2.59	-2.54	-2.49	-2.59	-
	p	0.79	0.80	0.55	0.63	0.88	0.93	0.36	-
	n	106	54	52	35	36	35	94	-
LVM	*Baseline vs. T$_{all}$:*								
	B	-2.29	-2.33	-2.24	-2.27	-2.36	-2.24	-2.30	-
	T$_{all}$	-2.40	-2.41	-2.38	-2.47	-2.40	-2.32	-2.47	-
	p	0.19	0.46	0.27	0.15	0.79	0.58	0.38	-
	n	83	45	38	28	29	26	37	-
	Baseline vs. T$_2$:								
	B	-2.30	-2.37	-2.21	-2.25	-2.38	-2.26	-2.49	-
	T$_2$	-2.33	-2.43	-2.20	-2.41	-2.33	-2.24	-2.42	-
	p	0.77	0.62	0.94	0.29	0.71	0.89	0.68	-
	n	76	43	33	26	26	24	28	-

Average deceleration level during conflict resolution (m/s^2):

		Overall	Gender		Age (years)			Speed (mph)	
			Male	Female	20-30	40-50	60-70	25-55	55+
LVD	*Baseline vs. T$_{all}$*:								
	B	-1.56	-1.58	-1.54	-1.58	-1.55	-1.56	-1.57	-
	T$_{all}$	-1.55	-1.56	-1.53	-1.56	-1.57	-1.51	-1.59	-
	p	0.54	0.48	0.83	0.67	0.70	0.31	0.71	-
	n	107	54	53	35	36	36	97	-
	Baseline vs. T$_2$:								
	B	-1.56	-1.58	-1.54	-1.58	-1.55	-1.55	-1.57	-
	T$_2$	-1.55	-1.57	-1.54	-1.58	-1.55	-1.53	-1.59	-
	p	0.73	0.58	0.99	0.93	0.92	0.68	0.74	-
	n	106	54	52	35	36	35	94	-
LVM	*Baseline vs. T$_{all}$*:								
	B	-1.51	-1.55	-1.47	-1.60	-1.47	-1.47	-1.59	-
	T$_{all}$	-1.62	-1.62	-1.63	-1.62	-1.63	-1.61	-1.74	-
	p	0.10	0.48	0.07	0.85	0.11	0.23	0.35	-
	n	83	45	38	28	29	26	37	-
	Baseline vs. T$_2$:								
	B	-1.52	-1.57	-1.45	-1.59	-1.49	-1.47	-1.74	-
	T$_2$	-1.55	-1.61	-1.48	-1.56	-1.60	-1.49	-1.68	-
	p	0.69	0.73	0.84	0.82	0.38	0.90	0.74	-
	n	76	43	33	26	26	24	28	-

Headway time at brake onset (seconds):

		Overall	Gender		Age (years)			Speed (mph)	
			Male	Female	20-30	40-50	60-70	25-55	55+
LVD	*Baseline vs. T$_{all}$*:								
	B	2.39	2.36	2.42	2.29	2.43	2.44	2.85	-
	T$_{all}$	2.39	2.42	2.36	2.30	2.36	2.52	2.74	-
	p	0.95	0.31	0.28	0.89	0.23	0.43	0.19	-
	n	105	54	51	34	36	35	94	-
	Baseline vs. T$_2$:								
	B	2.38	2.36	2.41	2.29	2.43	2.43	2.82	-
	T$_2$	2.41	2.45	2.36	2.26	2.36	2.61	2.88	-
	p	0.67	0.25	0.44	0.64	0.42	0.11	0.56	-
	n	104	54	50	34	36	34	91	-

Table 42. Average response intensities to lane-change conflicts

Maximum lateral acceleration (m/s^2):

		Overall	Gender		Age (years)			Speed (mph)	
			Male	Female	20-30	40-50	60-70	25-55	55+
LCR	*Baseline vs. T$_{all}$*:								
	B	0.78	0.70	0.93	0.90	0.69	0.74	-	0.79
	T$_{all}$	0.79	0.73	0.89	0.83	0.77	0.76	-	0.79
	p	0.93	0.67	0.78	0.62	0.24	0.75	-	0.98
	n	30	19	11	11	8	11	-	27
	Baseline vs. T$_2$:								
	B	0.87	0.76	1.00	0.98	-	-	-	0.88
	T$_2$	0.74	0.66	0.86	0.80	-	-	-	0.72
	p	0.25	0.42	0.46	0.38	-	-	-	0.18
	n	18	10	8	9	-	-	-	16
LCL	*Baseline vs. T$_{all}$*:								
	B	0.82	0.83	0.82	0.90	0.75	0.83	0.71	0.93
	T$_{all}$	0.82	0.82	0.81	0.83	0.86	0.73	0.91	0.80
	p	0.92	0.92	0.96	0.52	**0.05**	0.25	0.19	0.20
	n	52	29	23	19	21	12	13	39
	Baseline vs. T$_2$:								
	B	0.81	0.79	0.83	0.89	0.74	0.76	0.69	0.92
	T$_2$	0.82	0.80	0.84	0.81	0.89	0.69	0.87	0.79
	p	0.82	0.82	0.89	0.52	**0.02**	0.34	0.28	0.28
	n	43	22	21	18	17	8	8	32

Maximum lane bust distance (m):

		Overall	Gender		Age (years)			Speed (mph)	
			Male	Female	20-30	40-50	60-70	25-55	55+
LCL	*Baseline vs. T$_{all}$*:								
	B	0.30	0.28	-	-	0.21	-	-	0.36
	T$_{all}$	0.36	0.24	-	-	0.27	-	-	0.37
	p	0.46	0.74	-	-	0.54	-	-	0.87
	n	15	8	-	-	8	-	-	12
	Baseline vs. T$_2$:								
	B	0.40	-	-	-	-	-	-	-
	T$_2$	0.38	-	-	-	-	-	-	-
	p	0.84	-	-	-	-	-	-	-
	n	8	-	-	-	-	-	-	-

Table 43. Average number of road-departure conflicts per 100 miles driven

Departing straight road to the right (SDR):

	Overall	Gender		Age (years)			Light		Weather		Speed (mph)	
		Male	Female	20-30	40-50	60-70	Night	Day	Clear	Adverse	25-55	55+
Baseline vs. T_{all}:												
B	0.38	0.38	0.39	0.32	0.37	0.46	0.53	0.49	0.40	-	0.53	0.73
T_{all}	0.35	0.36	0.34	0.27	0.32	0.49	0.38	0.44	0.37	-	0.55	0.46
p	0.51	0.69	0.57	0.32	0.34	0.83	0.16	0.44	0.49	-	0.72	**0.03**
n	52	35	17	19	17	16	16	43	52	-	29	33
Baseline vs. T_2:												
B	0.42	0.40	0.47	0.38	0.37	0.50	0.53	0.49	0.44	-	0.57	0.79
T_2	0.41	0.36	0.55	0.36	0.34	0.53	0.37	0.52	0.43	-	0.71	0.56
p	0.86	0.40	0.51	0.80	0.67	0.82	0.17	0.68	0.89	-	0.20	0.09
n	42	31	11	12	16	14	12	34	42	-	21	25

Departing straight road to the left (SDL):

	Overall	Gender		Age (years)			Light		Weather		Speed (mph)	
		Male	Female	20-30	40-50	60-70	Night	Day	Clear	Adverse	25-55	55+
Baseline vs. T_{all}:												
B	0.77	0.76	0.78	0.83	0.89	0.57	1.01	0.89	0.79	-	1.00	1.34
T_{all}	0.74	0.78	0.69	0.72	0.80	0.70	0.80	0.82	0.75	-	0.88	1.12
p	0.60	0.72	0.27	0.15	0.39	0.14	0.21	0.31	0.46	-	0.13	0.09
n	86	50	36	31	29	26	32	78	84	-	66	62
Baseline vs. T_2:												
B	0.79	0.78	0.80	0.85	0.95	0.57	1.15	0.91	0.81	-	1.04	1.35
T_2	0.80	0.82	0.77	0.84	0.81	0.75	0.99	0.89	0.81	-	1.00	1.29
p	0.86	0.57	0.78	0.88	0.23	0.10	0.47	0.78	0.99	-	0.75	0.64
n	80	48	32	28	26	26	25	75	77	-	54	57

Table 44. Average response intensities to road-departure conflicts

Maximum lateral acceleration (m/s^2):

		Overall	Gender		Age (years)			Speed (mph)	
			Male	Female	20-30	40-50	60-70	25-55	55+
SDR	*Baseline vs. T$_{all}$*:								
	B	1.13	1.15	1.10	1.03	1.21	1.17	1.27	1.09
	T$_{all}$	1.16	1.07	1.33	1.17	1.18	1.11	1.17	1.12
	p	0.76	0.17	0.18	0.28	0.85	0.47	0.31	0.66
	n	52	35	17	19	17	16	29	33
	Baseline vs. T$_2$:								
	B	1.14	1.14	1.16	1.05	1.19	1.18	1.29	1.10
	T$_2$	1.08	1.05	1.16	0.96	1.17	1.08	1.15	0.98
	p	0.32	0.17	0.99	0.32	0.91	0.21	0.28	**0.03**
	n	42	31	11	12	16	14	21	25
SDL	*Baseline vs. T$_{all}$*:								
	B	1.23	1.23	1.23	1.29	1.23	1.16	1.35	1.13
	T$_{all}$	1.27	1.31	1.22	1.38	1.23	1.18	1.41	1.10
	p	0.55	0.38	0.85	0.54	0.98	0.73	0.55	0.47
	n	86	50	36	31	29	26	66	62
	Baseline vs. T$_2$:								
	B	1.24	1.24	1.25	1.30	1.26	1.16	1.40	1.12
	T$_2$	1.22	1.22	1.22	1.24	1.26	1.16	1.36	1.14
	p	0.63	0.76	0.71	0.45	0.98	0.96	0.74	0.74
	n	80	48	32	28	26	26	54	57

Average lane bust time (s):

		Overall	Gender		Age (years)			Speed (mph)	
			Male	Female	20-30	40-50	60-70	25-55	55+
CDR	*Baseline vs.* T_{all}:								
	B	2.03	2.17	1.84	1.94	2.04	2.13	2.08	2.05
	T_{all}	2.05	1.95	2.19	2.10	2.06	1.97	1.90	2.05
	p	0.87	0.17	**0.05**	0.40	0.93	0.49	0.37	1.00
	n	71	41	30	24	29	18	44	45
	Baseline vs. T_2:								
	B	1.92	2.07	1.70	1.92	1.80	2.12	1.97	2.00
	T_2	1.90	1.80	2.06	1.71	2.02	1.94	1.85	1.91
	p	0.89	0.11	0.10	0.41	0.27	0.55	0.63	0.56
	n	57	34	23	17	24	16	36	35
CDL	*Baseline vs.* T_{all}:								
	B	2.13	2.20	2.05	2.07	2.15	2.20	2.29	2.05
	T_{all}	2.03	1.96	2.11	1.88	1.99	2.24	2.16	1.97
	p	0.30	0.05	0.72	0.27	0.29	0.87	0.38	0.59
	n	86	48	38	31	30	25	72	59
	Baseline vs. T_2:								
	B	2.16	2.20	2.10	2.14	2.15	2.20	2.31	2.18
	T_2	2.04	1.92	2.20	1.90	1.93	2.33	2.07	1.97
	p	0.29	**0.04**	0.60	0.20	0.14	0.62	0.10	0.17
	n	83	48	35	29	29	25	67	47

Maximum Lane Bust Distance (m)

		Overall	Gender		Age (years)			Speed (mph)	
			Male	Female	20-30	40-50	60-70	25-55	55+
SDR	*Baseline vs. T_{all}:*								
	B	0.32	0.30	0.38	0.52	0.23	0.27	0.33	0.28
	T_{all}	0.26	0.27	0.22	0.25	0.25	0.29	0.26	0.23
	p	0.17	0.59	0.16	**0.01**	0.64	0.86	0.20	0.51
	n	31	23	8	8	13	10	18	14
	Baseline vs. T_2:								
	B	0.28	0.28	-	-	0.22	-	0.34	0.22
	T_2	0.28	0.28	-	-	0.25	-	0.28	0.27
	p	0.88	0.92	-	-	0.60	-	0.35	0.47
	n	21	16	-	-	11	-	12	9
SDL	*Baseline vs. T_{all}:*								
	B	0.27	0.27	0.28	0.29	0.30	0.23	0.29	0.26
	T_{all}	0.32	0.31	0.33	0.28	0.34	0.34	0.34	0.28
	p	**0.03**	0.07	0.21	0.95	0.16	**0.04**	0.23	0.44
	n	73	45	28	25	25	23	52	47
	Baseline vs. T_2:								
	B	0.27	0.27	0.27	0.29	0.29	0.20	0.30	0.25
	T_2	0.31	0.29	0.33	0.25	0.33	0.34	0.31	0.26
	p	0.24	0.52	0.32	0.33	0.39	0.06	0.73	0.71
	n	64	41	23	23	23	18	43	40
CDR	*Baseline vs. T_{all}:*								
	B	0.25	0.30	0.18	0.28	0.19	0.28	0.28	0.19
	T_{all}	0.27	0.24	0.31	0.29	0.23	0.30	0.25	0.24
	p	0.38	0.10	**0.00**	0.78	0.34	0.76	0.46	0.26
	n	48	27	21	17	19	12	35	23
	Baseline vs. T_2:								
	B	0.25	0.30	0.20	0.29	0.20	0.28	0.30	0.18
	T_2	0.25	0.21	0.30	0.26	0.21	0.30	0.24	0.24
	p	0.96	0.08	**0.02**	0.61	0.90	0.76	0.23	0.22
	n	38	21	17	14	15	9	27	18
CDL	*Baseline vs. T_{all}:*								
	B	0.27	0.27	0.28	0.30	0.25	0.26	0.33	0.22
	T_{all}	0.28	0.29	0.26	0.25	0.29	0.30	0.33	0.24
	p	0.88	0.43	0.53	0.23	0.17	0.43	0.91	0.53
	n	79	44	35	30	27	22	55	48
	Baseline vs. T_2:								
	B	0.28	0.27	0.29	0.31	0.26	0.26	0.31	0.22
	T_2	0.28	0.29	0.26	0.26	0.28	0.31	0.30	0.23
	p	0.87	0.47	0.50	0.31	0.40	0.25	0.80	0.76
	n	68	44	24	25	24	19	44	35

Table 45. Average number of conflicts where drivers approached curves at excessive speed per 100 miles driven

	Overall	Gender		Age (years)			Light		Weather		Speed (mph)		
		Male	Female	20-30	40-50	60-70	Night	Day	Clear	Adverse	<25	25-55	55+
Baseline vs. T_{all}:													
B	1.10	1.28	0.77	1.27	1.26	0.58	2.30	1.21	1.14	-	5.44	2.64	0.59
T_{all}	1.07	1.26	0.74	1.15	1.30	0.59	1.90	1.18	1.11	-	4.72	2.48	0.55
p	0.70	0.79	0.77	0.47	0.74	0.86	0.32	0.79	0.72	-	0.36	0.37	0.79
n	75	48	27	29	28	18	32	69	74	-	18	73	18
Baseline vs. T_2:													
B	1.12	1.28	0.81	1.27	1.33	0.58	2.29	1.25	1.17	-	5.47	2.69	0.63
T_2	1.14	1.32	0.80	1.13	1.52	0.62	2.04	1.31	1.19	-	6.65	2.77	0.73
p	0.86	0.80	0.95	0.50	0.35	0.63	0.58	0.69	0.87	-	0.48	0.73	0.51
n	73	48	25	29	26	18	28	64	72	-	14	71	14

Table 46. Average delta speed at CPOI during curve-speed conflicts

	Overall	Gender		Age (years)			Speed (mph)		
		Male	Female	20-30	40-50	60-70	<25	25-55	55+
Baseline vs. T_{all}:									
B	2.24	2.36	2.04	2.27	2.28	2.15	1.83	2.23	3.39
T_{all}	2.31	2.44	2.07	2.41	2.36	2.06	1.76	2.33	3.91
P	0.24	0.24	0.72	0.15	0.30	0.43	0.44	0.11	0.24
N	75	48	27	29	28	18	18	73	18
Baseline vs. T_2:									
B	2.26	2.36	2.07	2.27	2.33	2.15	1.84	2.25	3.52
T_2	2.32	2.42	2.14	2.44	2.38	2.05	1.69	2.35	3.97
P	0.32	0.45	0.52	0.13	0.56	0.37	0.08	0.16	0.34
N	73	48	25	29	26	18	14	71	14

Appendix H: Near Crash Thresholds by Conflict Type

Table 47. Near crash thresholds by conflict type

Conflict Type		Variable	Value	Number of Near Crashes
Rear-end		POV is moving		370
		Min TTC	< 3 s	
		Max deceleration	> 4.0 m/s	
		Brake duration	> 0.5 s	
Curve speed		Max lateral acceleration	> 3.5 m/s²	274
		Speed reduction at tightest point of curve	≥ 3 m/s	
		OR		
		Max lateral acceleration	> 4.5 m/s²	188
		Speed reduction at tightest point of curve	< 3 m/s	
Lane change	Straight road	No lane excursion		38
		Max lateral acceleration	≥ 1.0 m/s²	
		OR		
		Maximum lane excursion	0.1 m - 0.3 m	46
		Max lateral acceleration	≥ 0.75 m/s²	
		OR		
		Maximum lane excursion	0.3 m - 0.9 m	36
		Max lateral acceleration	≥ 0.0 m/s²	
	Depart to outside of curve	No lane excursion		90
		Max lateral acceleration	≥ 0.5 m/s²	
		Normalized relative acceleration	> 0.25	
		OR		
		Maximum lane excursion	0.1 m - 0.9 m	21
		Max lateral acceleration	≥ 0.0 m/s²	
		Normalized relative acceleration	> 0.25	
	Depart to inside of curve	Maximum lane excursion	0.1 m - 0.9 m	20
		Max lateral acceleration	≥ 0.0 m/s²	
		Normalized relative acceleration	> 0.75	
		OR		
		No lane excursion		66
		Max lateral acceleration	≥ 0.0 m/s²	
		Normalized relative acceleration	> 0.75	

Conflict Type		Variable	Value	Number of Near Crashes
Road Departure	Straight road	Maximum lane excursion	0.1 m - 0.3 m	130
		Max lateral acceleration	≥ 1.5 m/s^2	
		OR		
		Maximum lane excursion	0.3 m - 0.9 m	287
		Max lateral acceleration	≥ 1.0 m/s^2	
	Depart to outside of curve	Maximum lane excursion	0.1 m - 0.9 m	250
		Max lateral acceleration	≥ 1.0 m/s^2	
		Normalized relative acceleration	> 0.25	
	Depart to inside of curve	Maximum lane excursion	0.1 m - 0.9 m	215
		Max lateral acceleration	≥ 2.5 m/s^2	
		Normalized relative acceleration	> 2.25	

Appendix I: Post-Drive Survey Mapping to Acceptance Objectives

			Data source				
			Questionnaire	Debrief	Focus Group	Video Analysis	Numerical	Pre-Drive Questionnaires

			Questionnaire	Debrief	Focus Group	Video Analysis	Numerical	Pre-Drive Questionnaires
Objectives	**1. Ease of use**	1.1 Usability of the warnings	9, 14, 37, 43, 44, 45, 46, 47		4.1, 4.2			
		1.2 Usability of the DVI						
		i Usability of warning modalities	10, 17, 19, 20, 22, 23, 25, 26, 28		19.2, 7.0			
		ii Usability of controls/Display	38, 39, 40		14.1			
		1.3 Understanding of the warnings	15, 16, 18, 21, 24, 27	X				
		1.4 Demands on drivers	8					
		1.5 Warning patterns	9, 29					
	2.2 Perceived usefulness	2.1 Usefulness of warnings	4, 5, 11, 30, 33, 34, 35, 36, 37, 43, 44, 45, 46, 47	X	3.1, 17.0, 18.0			
		2.2 Safety						
		i increase in driving safety due to IVBSS	6					
		ii Increase in awareness of surroundings due to IVBSS	7					
		2.3 Tolerance of nuisance warnings						
		i Annoyance with nuisance warnings	14, 31, 32					
		ii Assessment of impact of nuisance warnings on driving		X	16.0			
	2.3 Ease of learning	3.1 Utility of instruction/training						
		i Time required to become familiar with operation of IVBSS			14.2, 15.1			
		ii Assessment of ability to use IVBSS correctly						
		3.2 Comprehension			4.2, 5.1, 19.1			
	2.4 Advocacy	4.1 Willingness to use IVBSS	41		8.0, 9.0, 10.0, 11.0, 12.0, 13.0			
		4.2 Interest in purchasing	42, 48, 49, 50, 51, 52		5.3, 6.0			
		4.2 Resistance to new technology			20.0			
	2.5 Driving performance	5.1 Control Input						
		i Snooze button use (frequency/conditions)	39					
		ii Volume use	40					
		5.2 Vigilance				X		
		5.3 Unintended consequences	12, 13					
Independent Variables	**Demographic/ Driving History**	Age						X
		Gender						X
		Years with Driver's license						X
		Driving record						X
		Annual mileage						X
		DBQ/DSQ						X
		Prior experience with advanced safety systems						X
	IVBSS Experience	LDW Availability					X	
		Driving Patterns					X	
		Intensity of Experience						
		Prob of a conflict					X	
		Prob of an alert					X	
		Prob of a cofflict/alert					X	
		Prob of alert/conflict					X	
		Occurrence of near crash			3.2	X		
		System integration (warning clusters)					X	

DOT HS 811 516
October 2011

U.S. Department
of Transportation
**National Highway
Traffic Safety
Administration**